Advanced Manufacturing Processes

The field of manufacturing over the years has seen the introduction of new and innovative technologies for enhancing output, increasing quality and reducing material inventory. This textbook discusses fundamental concepts, principles, technologies and applications of advanced manufacturing processes. It comprehensively discusses key manufacturing technologies including reconfigurable manufacturing processes, computer integrated manufacturing processes, agile manufacturing processes, cellular manufacturing processes, rapid prototyping and flexible manufacturing processes.

This book:

- Discusses rapid prototyping techniques in detail.
- Covers automation systems and their techniques.
- Examines issues related to reconfigurable manufacturing processes, flexible manufacturing processes and agile manufacturing processes.
- Discusses actuators and sensors, industrial robots, lean production, comparison of lean and agile and rapid prototyping techniques.
- Covers matrix approach techniques to understand the barriers index value.

The textbook is primarily written for senior undergraduate and graduate students in the field of mechanical engineering, industrial and production engineering.

Advanced Manufacturing Processes

Vasdev Malhotra

CRC Press
Taylor & Francis Group
Boca Raton London New York

CRC Press is an imprint of the
Taylor & Francis Group, an **Informa** business

Designed cover image: Shutterstock

First edition published 2024
by CRC Press
2385 NW Executive Center Drive, Suite 320, Boca Raton FL 33431

and by CRC Press
4 Park Square, Milton Park, Abingdon, Oxon, OX14 4RN

CRC Press is an imprint of Taylor & Francis Group, LLC

ISBN: 9780367750565 (hbk)
ISBN: 9781032759593 (pbk)
ISBN: 9781003476375 (ebk)

DOI: 10.1201/9781003476375

Typeset in Times
by Newgen Publishing UK

Contents

About the Author

Prof. (Dr.) Vasdev Malhotra, an eminent figure in Mechanical Engineering, illuminates the corridors of J.C. Bose University of Science and Technology (State Government University, Faridabad, Haryana), with a career spanning over two decades. His influence transcends academic boundaries.

A luminary in the field, Dr. Malhotra boasts a remarkable journey—a Governor Chancellor nominee of the University of Haryana, esteemed advisor to UPSC and a revered expert within AICTE. In 2019, his outstanding contributions were acknowledged with the Bharat Ratna Dr. Radhakrishnan Gold Medal Award by the Global Economic Progress and Research Association.

His educational foundation is robust as his professional accolades, holding degrees from prestigious institutions like NIT Kurukshetra, PTU Jalandhar and MDU Rohtak. Dr. Malhotra's academic repertoire spans over 106 research papers, notably in esteemed international journals indexed by SCI and Scopus.

As a mentor, he guided numerous M.Tech and B.Tech projects and mentored Ph.D. students. His literary contributions include chapters in globally recognized publications such as IGI Global Publication, Australia.

Beyond research, Prof. Malhotra actively engages in the scholarly community, completing projects of DST and AICTE and contributing to industrial consultancy projects. His administrative roles and commitment to social initiatives underscores his dedication to holistic education and societal development.

A multi-dimensional contributor, Prof. Malhotra serves on advisory committees, holds membership in professional societies and reviews international journals and conferences globally, and has delivered keynote lectures in international conferences held in Singapore and Dubai.

Prof. (Dr.) Vasdev Malhotra emerges not just as a stalwart in Mechanical Engineering but as an exemplary figure, leaving an indelible mark on education, engineering and societal development.

1 Reconfigurable Manufacturing Systems

1.1 INTRODUCTION

Dedicated machine tools and controllers were widely used among manufacturing enterprises before the first numerically controlled machine was invented. During this period most machine tools and controllers were purely mechanical or electromechanical systems. The major disadvantage of these systems was that each machine tool and controller was tailored for a special product. The function of a dedicated machine tool controller could not be upgraded without great difficulty. As customer demands for different products changed over time, manufacturing enterprises often had to replace dedicated machine tools and controllers to accommodate this demand. Table 1.1 (see page xx) shows the drawbacks of dedicated manufacturing and flexible manufacturing.

These technologies have drastically changed the way parts are designed and manufactured. CNC machines and controllers have brought many benefits into manufacturing systems by improving product quality, product accuracy and machine control accuracy. Reconfigurable manufacturing system came into existence in 1998. The inventor noted the deficiencies of existing computer numerical machine tools and controllers, which included a lack of interchangeability. Reconfigurable manufacturing systems may be defined as a manufacturing system designed at the outset for rapid changes in hardware and software components, so as to quickly adjust production capacity within a part family in response to sudden changes in market. An RMS is expected to be able to rapidly adjust to new circumstances by rearranging its hardware and software components in order to accommodate not only the production of a variety of products, but also the new product introduction within each family.

1.2 BARRIERS TO RMS

On the basis of literature reviews and experiences of manufacturing managers and academicians, it has been found that the design and implementation process of RMS is a very complicated task. The barriers affecting the implementation of reconfigurable manufacturing system are discussed below.

DOI: 10.1201/9781003476375-1

1.2.1 DEVELOPMENT OF DESIGN METHODOLOGY

The development of a mathematical framework for synthesis of reconfigurable machines and their validation is a major barrier. So the development of a mathematical theory for synthesis of reconfigurable machines, which includes development of a formal and a unified representation scheme for mechanical function of modules, is very difficult.

1.2.2 DIFFICULT INTERFACES

It is not so easy to assemble machines modularly accurately enough to meet the accuracy requirement of machine tools. The interfaces between the modules to be assembled must be standardized and accurately manufactured. The methodologies must be developed to rapidly measure and adjust the alignment of modules.

1.2.3 MODULE ECONOMY

Most movable and drive units are supplied with electricity and are connected to the controller by wires. Each module should be autonomous and independently work-able. This is a big barrier for reconfigurable machine units.

1.2.4 RECONFIGURATION OF CONTROLLER ARCHITECTURE

Reconfiguration of controller architecture is not an easy task, because to allow recon-figuration, machine tools should be integrated which will be done by a software module.

1.2.5 DIFFICULT CONTROL OF RECONFIGURABLE MACHINE TOOLS WITH MULTIPLE TOOLS

Multiple tools are required for reconfiguration of a machine, However, movement control of multiple tools in a non-orthogonal direction is difficult, but nowadays different strategies are developing for interpolation and control of RMT with axes in non-orthogonal configurations.

1.2.6 INTEGRATION OF HETEROGENEOUS SOFTWARE AND HARDWARE COMPONENTS

The integration of software with hardware components is difficult because the inte-gration of heterogeneous software and hardware components, which are developed by different vendors at different times, will require special software and electrical interfaces.

1.2.7 DIFFICULTY IN AXES LOCATION

Location of axes in the reconfigurable machine is not an easy task. If the proper loca-tion of the axes is not attained by the machine, then it will result in loss of accuracy.

1.2.8 RECONFIGURATION OF CONTROL SYSTEM

The control systems are reconfigured by using advanced software and hardware modules for development of an open-ended control architecture. The selection of modules are directly influenced by the electrical components.

1.2.9 EXPENSIVE TOOLING

To optimize the cost of a manufacturing process a costing model is required, but a reconfigurable machine requires expensive tooling. Sometimes tools from the two to six sets which are generally required may cost as much as the machines.

1.2.10 DIFFICULT VARIETY HANDLING

The large number of individual products in RMS inevitably leads to a high variety of material items such as raw materials and work in progress. Since their fulfillment is in the control focus of RMS, it is essential to capture the high variety of material and end products in small models.

1.2.11 CONTROLS OF PROCESS VARIATIONS

The product variety is associated with diverse design specifications as design changes lead to many changes in the processes of producing material items and end products. Such changeovers are reflected as variations in machines, operations and processes. Therefore, the system model should be able to capture and reflect the variations, and this process is very difficult.

1.2.12 COMPLEX SYSTEM

In RMS, a number of processes can be used to produce one product. Such processes relate to different configurations of different machines. In practice, only one process is adopted to produce a product. In accordance with various products to be produced in the same time periods using identical resources, the proper machines and processes must be selected from multiple alternatives.

1.2.13 CONSTRAINT SATISFACTION

In RMS, many constraints can be observed. These constraints are inherent in the selection of machines and operations and are associated with specific design and machine capabilities. To deal with these constraints in the modeling of RMS is a difficult task.

1.2.14 SELECTION OF MACHINE MODULES

In RMS, a number of processes may be used to produce one product. Such processes relate to different configurations of different machines. In practice, only one process is adopted to produce a product. In accordance with various products to be produced in the same time periods using identical resources, the proper machines and processes must be selected from multiple alternatives. This selection should contribute to the improvement of certain system performance attributes. Hence, system models should facilitate decision-making in selecting machines and processes that are difficult.

1.3   ENABLERS AFFECTING RECONFIGURABLE MANUFACTURING SYSTEMS

Systems design helps understand some general design rules, which increase flexibility and reconfigurablity. There are a number of factors to be decided on when designing a flexible and reconfigurable production system. From literature reviews some of the factors which affect reconfigurable system design are discussed below.

1.3.1   IN-HOUSE PRODUCTION

Regarding flexibility, some companies believe that internal manufacturing gives better control. Still, when it comes to growth and long-term changes internal manufacturing can be an obstacle. This indicates that keeping manufacturing in-house can increase flexibility and outsourcing can increase reconfigurablity.

1.3.2   COOPERATION

This might reduce the ability for reconfiguration of the supply chain if the possibility to change suppliers decreases. Still, this will be leveraged with improved cooperation in product and process development, which will improve capabilities to release new products.

1.3.3   AUTONOMY

By intensive controlling it is possible to make lots of re-planning and adjustment to the process. At the same time, autonomy lets the employees' creativity and willingness to run operations increase, which in the long run is a basis for motivation.

1.3.4   ORDINARY EMPLOYEES

Ordinary employees are the basis for future development and internal improvements. Temporary employees can be an effective instrument in handling a volume increase. As future development is vital for reconfigurablity and motivation from the organization, this development is very important. It is suspected that a temporary employee does not have the same motivation for long-term development. Still, hiring temporary employees could be very effective at specific times, to handle temporary volume variations.

1.3.5 SHORT-TERM HORIZONS

It seems that the economy of today becomes more and more short-term. Thus, the demands from shareholders will always push for short-term decisions and therefore it is important for a company to focus on a longer-term horizon.

1.3.6 AUTOMATIC MANUFACTURING

Automatic manufacturing equipment gives advantages in quality, work-environment and capacity. Still, highly automated machinery is not very flexible in terms of redesigning and reconfiguring the system.

1.3.7 STANDARD PRODUCTS

From a manufacturing point of view, a standard product is preferable. Still, customization is a base for competition in a turbulent environment. The principles and practices of modularization and systems design are central and common in many future production systems philosophies.

1.4 VARIOUS ISSUES AFFECTING RECONFIGURABLE MANUFACTURING SYSTEMS

The following are the various issues affecting reconfigurable manufacturing systems.

1.4.1 DESIGN OF RECONFIGURABLE MACHINES

The concept of reconfigurable machine tools (RMTs) appeared in the early 1990s as a particular trend that evolved directly from the concept of FMS. Machine tools available in today's markets have been designed for flexibility in terms of the types of processes and the geometric complexity of the product that can be manufactured. This flexibility is provided at a high cost, given that only a very limited subset of the machine tool's capabilities is used at any given instant. Clearly, the user must pay for any unused capacity. The concept of reconfigurability imposes very specific constraints on the structural design of a machine. Quick design and realization are important components of this technology and as a consequence the dimensions that must be incorporated in RMTs are:

- Ease of assembly and integration with monitoring, actuation systems.
- Design for structural integrity and dynamic stiffness of machine tool.
- Design for rapid ramp up systems for rapid diagnosis.

1.4.2 MANUFACTURING PROCESS AND SIMULATION MACHINES

Machine intelligence is greatly enhanced through the incorporation of manufacturing process models into the controllers. There is extensive research on modeling of manufacturing processes. The current trend is gradually incorporating these manufacturing process models into machine controllers, together with additional machine sensors. With process-oriented controllers and associated programming systems, there are multiple benefits in the operation of machine tools, for example: increased machine productivity through adaptive control, improved machine reliability through avoidance of process instability.

1.4.3   MICRO ELECTRO–MECHANICAL (MEMS) DEVICES FOR SENSORS

The design of new reconfigurable machines will therefore depend on intelligent use of sensor and actuator devices to reduce costs without compromising capabilities. In that sense, MEMS-based sensors and actuators have an enormous potential. Microsystems involve both electronic and non-electronic elements, and perform functions that can include sensing, signal processing, actuation, control and display. MEMS are fabricated either by bulk micro-machining or surface micro-machining. There are many advantages of MEMS devices compared with the elements they replace. For example, they can be so small that hundreds of them can fit in the same space as one single macro-device that performs the same function.

1.4.4   INTEGRATED MACHINE DEVELOPMENT BASED ON SCIENTIFIC ENGINEERING

In machine development, a synergetic combination of mechanics, electronics, software and computing requires that individuals possess a multi-disciplinary understanding of relevant scientific and engineering principles. This individual knowledge must be sufficiently comprehensive to be able to create an innovative combination, which makes up a mechatronic solution. To support this mechatronic solution, appropriate design methods and tools backed by computer design facilities are needed. The individual software tools for mechanical, electronic and software engineering are all available, and there has been considerable progress in bringing them together to provide a coherent simulation environment by which the performance of a given mechatronic system can be assessed.

1.4.5   EFFECTIVE NETWORK CONTROL SYSTEM SOLUTIONS

To achieve effective network control system solutions for RMS control systems a number of issues must be addressed, in addition to the performance analysis. One important issue involves the utilization of smart devices which are smart sensors, smart actuators and networked controllers. Smart sensors or actuators have three major features: intelligence, communication ability and data acquisition or actuation, respectively. Besides network-capable application processors, the major functionalities of networked controllers are to analyze the sensor data, make decisions and give commands to actuation devices. The control algorithms should handle decentralized information analysis as well as the traditional centralized cases.

1.5 COMPONENTS OF RECONFIGURABLE MANUFACTURING SYSTEMS

Reconfigurable manufacturing systems have two important components:

- Reconfigurable machine tool;
- Reconfigurable controller.

1.5.1 RECONFIGURABLE MACHINE TOOL

The importance of reconfigurable manufacturing systems is that the structure of the system as well as of its machines and control can be rapidly changed with related market changes. A major component of a reconfigurable manufacturing system is RMT (i.e., reconfigurable machine tool).

By contrast to conventional CNCs that are general purpose machines, RMT are designed for a specific customized range of operational requirements and may be cost effectively converted when the requirements change. RMT are designed to produce a specific set of features for a specific range of cycle times. Some operational requirements will be constant over the lifetime of the machining system. The aim of the reconfigurable machine tool is to cope with various changes in the products to be manufactured. The following possible changes must be taken into consideration:

- Work piece size;
- Part geometry and complexity;
- Production volume and production rates;
- Required processes;
- Accuracy requirement in terms of geometrical accuracy, surface quality;
- Material property such as material hardness, etc.

Reconfiguration requirements introduce several new challenges for RMT controllers. The first challenge is the reconfiguration of the controller architecture that is required when the physical machine tools are reconfigured, or new technology is integrated. Another challenge is the control of RMT with multiple tools working independently and RMT with axes in a non-orthogonal configuration. The other important challenge is the integration of heterogeneous software and hardware components. The design of RMT requires a broad knowledge of machine design, machine tool design, kinematics modeling and dynamics analysis. There is no comprehensive theory or design methodology that is directly applicable to RMT design. The concept of reconfiguration has been used in related fields including fixture design, assembly system design and reconfigurable robots. Some typical motion types are defined as follows.

1.5.2 RMT TOOL MACHINING MOTION

RMT tool machining motion collects all reconfigurable machine tool motions, which are involved in the same machining feature (hole, slot, etc.). All tool point motions have the same directions (x-axis, y-axis, etc.) and perform the same machining task.

1.5.3   RMT TOOL MACHINING MOTION FAMILY (TMMF)

An RMT TMMF is a collection of a series of machining motions among different work pieces, which can share the same motion types. According to reconfigurable machines, tools have four important modules to develop reconfigurable machine tool control systems as follows:

1. Automatic part transfer system control module;
2. Automatic part clamping rotating system control module;
3. Automatic part lifting system control module;
4. Automatic tool changing system control module.

1.5.4   RECONFIGURABLE CONTROLLER

To control a particular machine, any machine-specific functions or classes currently must be designed and built into a reconfigurable controller. The reconfigurable controller becomes unchangeable at run-time for controlling different machines. For instance, a reconfigurable controller can control a three-axis tabletop mill and a five-axis mill. To allow reconfiguration of the motion planner and servo controller necessary for controlling different mechanisms, some additions and modifications must be made to the reconfigurable controller. The configuration system is directly interfaced so that it can receive configuration commands from this interface. After the configuration system finishes all of these configuration processes, the reconfigurable controller is dynamically reconfigured for a particular mechanism.

1.6   RECONFIGURABLE MANUFACTURING SYSTEM CAPABILITIES

The reconfigurable capabilities could be divided into four different main dimensions:

- Product related change capabilities.
- Change competency within operations.
- Cooperation internally and externally.
- People, knowledge and creativity.

1.6.1   PRODUCT-RELATED CHANGE CAPABILITIES

Design for manufacturing is the most important area that plays a vital role to improve the product related capabilities, and if the products are designed for assembly this will make the system more productive. The products could also be designed for reconfigurablity in the sense of using similar materials, redundancy and efficient change-over. Information technology is also an important area, which helps to improve the product-related change capabilities.

1.6.2 CHANGE COMPETENCIES WITHIN OPERATIONS

In the decision process the aspects of generating alternatives and evaluating are vital for successful decisions. Also, a condition for any decision process is the ability to generate and collect information. The gathering of information, generation of alternatives and evaluation depends on which problem is being investigated. Thus, creativity and innovations are central to formulating alternatives and solutions.

1.6.3 COOPERATION INTERNALLY AND EXTERNALLY

Product design and methods as design for assembly require good cooperation between design and manufacturing. Several methodologies of concurrent engineering have been developed so that this can be used to improve cooperation. As the production system spans over several actors it becomes more and more important to cooperate with other companies.

1.6.4 KNOWLEDGE, CREATIVITY AND PEOPLE

The last defined dimension of reconfigurable capabilities is people, knowledge and creativity. Numerous ways to develop these capabilities exist, mainly related to management issues. Motivation can be developed by good management and is in itself a large area for research.

1.7 RECONFIGURABLE MANUFACTURING SYSTEM CHALLENGES

The concept of reconfiguration has sparked interest in the academic and industrial communities. It has encouraged active research into supportive areas that are proving very beneficial to existing manufacturing systems, e.g., in the areas of process and production planning, modular interfaces and the like. The challenges of reconfigurable manufacturing systems are:

1. Measures for changeability, flexibility, adaptability, responsiveness, reconfigurability and their relationships.
2. The hardware and software enabling technologies.
3. Reconfigurable logical support systems, such as logistics, production planning and control, process planning, tooling and fixtures.
4. Balance of hard and soft capacity and functionality scalability options.
5. Design of machines, systems and controls for flexibility, changeability and reconfiguration, and integration with current systems and software.
6. Models to determine adequate levels of changeability, flexibility and reconfigurability required for different applications.
7. Appropriate capacity scalability policies.
8. Appropriate frequency of change or reconfiguration.
9. Smooth and optimal systems transition and changeover.

10. Changeability and reconfiguration dependent quality factors, including human–related issues.
11. Complexity measurement, reduction and management techniques.
12. The use of group technology to capitalize on commonality and standardization of parts, operation sequences, product structure, platforms, engineering.

1.8 RECONFIGURABLE MANUFACTURING SYSTEMS CHARACTERISTICS

Based on literature review, various types of characteristics of Reconfigurable Manufacturing System follow.

1.8.1 MODULARITY

In a reconfigurable manufacturing system, many components are typically modular. The modular components can be replaced or upgraded to better suit new applications. The modules are easier to maintain and update, thereby lowering lifecycle costs of systems. New calibration algorithms can be readily integrated into the machine controller, resulting in a system with greater accuracy.

1.8.2 INTEGRABILITY

To provide the ability to integrate modules rapidly, a set of mechanical and control interfaces which enable integration and communication are needed. At the machine level, axes of motions and spindles can be integrated to form machines. Integration rules which allow machine designers to relate clusters of part features and their corresponding machining operations to machine modules are required.

1.8.3 CUSTOMIZATION

This characteristic drastically distinguishes RMS from flexible manufacturing systems and allows a reduction in investment cost. It enables the design of a system for the production of a part family, rather than a single part. The parts that have similar geometric features and shapes, the same level of tolerances, require the same processes and are within the same range of cost.

1.8.4 CONVERTIBILITY

It may require switching spindles on a milling machine (e.g., from a low-torque high-speed spindle for aluminum to a high-torque low-speed spindle for titanium), or manual adjustment of passive degrees-of-freedom changes when switching production between two members of the part family within a given day. The system conversion at this daily level must be carried out quickly to be effective.

1.8.5 SCALABILITY

It may require at the machine level adding spindles to a machine to increase its productivity and at the system level changing part routing or adding machines to expand the overall system capacity.

1.8.6 DIAGNOSABILITY

Diagnosability has two aspects: detecting machine failure and detecting unacceptable part quality. As production systems are made more reconfigurable, and their layouts are modified more frequently, it becomes essential to rapidly tune the newly reconfigured system so that it produces quality parts.

1.8.7 DYNAMIC BEHAVIOR

Adaptability of manufacturing systems mainly refers to activities reactively adopted by a manufacturing system to adapt to environmental changes. High adaptability of a reconfigurable manufacturing system permits quick and cost-effective response to changes.

1.8.8 INTEGRATED INFORMATION

Manufacturing systems are able to select the necessary data from the data-overloaded environment and to filter the right information. This will be a major core competence of a system to guarantee it survives in a fast-changing and fast-moving environment. As well as the management of a huge amount of data, there is also the necessity for systems to ensure that the right people get the necessary information at a proper quality, and also in time. Information integration meets requirements of communication and information sharing among autonomous units. Therefore, information integration is strongly related to the intelligence and decision-making capabilities of reconfigurable organization units.

1.8.9 ORIENTED INNOVATION

Innovations will be one of the best characteristics supporting manufacturing reconfigurablity concerning products, processes, organizations and markets. There may be different kinds of innovations, such as technological or social innovations. Social innovations influence the behavior of the people within manufacturing systems and create an environment where people's imagination and creativity are encouraged. On the other hand, technology innovations are the activities that create new market demands.

1.8.10 ENVIRONMENTAL CONSCIOUSNESS

Reconfigurable manufacturing systems have to be environmentally conscious. It takes the ecological concerns in the whole of manufacturing systems and products into account. Reconfigurablity of manufacturing systems and products will reduce and avoid negative impacts on the environment to ensure a sustainable development.

1.8.11 COMPETITION AND COOPERATION

Competition between manufacturing systems can be identified as another enabler to force the reconfigurablity of systems. It is one factor that leads internal adaptation to be more competitive. Cooperation can be seen as a characteristic that causes internal changes but mainly influences the market and the behavior of competitors.

THEORETICAL QUESTIONS

Question 1. What is a reconfigurable manufacturing system?

Question 2. Explain a dedicated manufacturing system with a suitable example.

Question 3. Differentiate between a reconfigurable manufacturing system, a dedicated manufacturing system and a flexible manufacturing system with an example.

Question 4. Explain the barriers affecting the design of a reconfigurable manufacturing system.

Question 5. Explain the issues related to a reconfigurable manufacturing system with examples.

Question 6. Explain the characteristics of reconfigurable manufacturing systems with suitable examples.

Question 7. Explain the factors affecting the implementation of reconfigurable manufacturing systems.

Question 8. Discuss the challenges faced by the industry expert during the implementation of a reconfigurable system in the industry.

Question 9. Define the following terms:
1. RMT tool machining motion;
2. RMT tool machining motion family.

Question 10. Explain the components of a reconfigurable manufacturing system with suitable applications.

Question 11. Explain the merits, demerits and applications of reconfigurable manufacturing systems.

Question 12. Explain the working of a reconfigurable controller with suitable merits and demerits, along with limitations.

TABLE 1.1
Weaknesses of Conventional Manufacturing Systems

DMS	FMS
Not flexible	Expensive
For a single part	Machine focus
Fixed capacity	Low throughput

MULTIPLE CHOICE QUESTIONS

Question 1. Tool life is measured by the:
   a) Number of pieces machined between tool sharpening;
   b) Time the tool is in contact with the job;
   c) Volume of material removed between tool sharpening;
   d) All of the above.

Question 2. Reconfigurability denotes the:
   a) Reconfigurable computing capability of a system;
   b) Machinability of a system;
   c) Both (a) and (b);
   d) None of the above.

Question 3. The following are the components of a reconfigurable manufacturing system:
   a) Reconfigurable machine tool;
   b) Reconfigurable controller;
   c) Both (a) and (b);
   d) None of the above.

Question 4. Integrability stands for:
   a) The ability to integrate modules rapidly;
   b) Related to machinability;
   c) Ability to increase efficiency;
   d) Ability to increase productivity.

Question 5. The following are the capabilities of a reconfigurable manufacturing system:
   a) Cooperation internally and externally;
   b) People, knowledge and creativity;
   c) Both (a) and (b);
   d) None of the above.

Question 6. The concept of reconfigurable machine tools (RMTs) appeared in:
   a) 1990;
   b) 1991;
   c) 1992;
   d) 1993.

Question 7. ------------------ is the most important area that plays a vital role in improving product related capabilities:
   a) Design for production;
   b) Design for manufacturing;
   c) Design for efficiency;
   d) Design for variables.

Question 8. TMMF is defined as:
   a) Tool machine motion family;
   b) Tool method motion family;
   c) Tool motion method family;
   d) None of the above.

Question 9. RMT stands for:
    a)   Reconfigurable machine tool;
    b)   Reconfigurable method tool;
    c)   Both (a) and (b);
    d)   None of the above.

Question 10. MEMS stands for:
    a)   Micro electro-mechanical systems;
    b)   Micro mechanical electro systems;
    c)   Both (a) and (b);
    d)   None of the above.

Question 11. The automatic part transfer system control module is the module to develop:
    a)   Efficiency tool;
    b)   Reconfigurable machine tool;
    c)   Integrity tool;
    d)   Productivity tool.

Question 12. A reconfigurable manufacturing system:
    a)   Has the ability to arrange integrity tools;
    b)   Is not able to rearrange the manufacturing elements;
    c)   Has the ability to rearrange the manufacturing elements;
    d)   None of the above.

Question 13. Oriented innovations are the:
    a)   Types of manufacturing systems;
    b)   Characteristics of manufacturing systems;
    c)   Both (a) and (b);
    d)   None of the above.

Question 14. -------------------------- is the challenge of a reconfigurable manufacturing system to:
    a)   Appropriate frequency of change;
    b)   Measure changeability;
    c)   Reconfigure logical support system;
    d)   All of the above.

Question 15. Ideal reconfigurable manufacturing systems possess how many core RMS characteristics:
    a)   Eight;
    b)   Six;
    c)   Four;
    d)   Two.

Question 16. RMS science base may be utilized to maximize system productivity with the smallest possible number of machines.
    a)   RMS science base;
    b)   RMS production base;
    c)   Both (a) and (b);
    d)   None of the above.

Question 17. RMS characteristics are:
   a)   Modularity;
   b)   Integrability;
   c)   Customization;
   d)   All of the above.
Question 18. RMS increases the --------------------- of manufacturing systems:
   a)   Speed of responsiveness;
   b)   Speed of scalability;
   c)   Both (a) and (b);
   d)   None of the above.
Question 19. Reconfigurable machine tools (RMT) were invented in:
   a)   2000;
   b)   1999;
   c)   2001;
   d)   2002.
Question 20. Integrability is the ability to integrate ----------- rapidly:
   a)   Modules;
   b)   Customization;
   c)   Both (a) and (b);
   d)   None of the above.
Question 21. Diagnosability has two aspects:
   a)   Detecting speed of failure and detecting acceptable part quality;
   b)   Detecting machine failure and detecting unacceptable part quality;
   c)   Both (a) and (b);
   d)   None of the above.
Question 22. The RMS possesses hardware and software capabilities to cost-effectively respond to which unpredictable events:
   a)   Hardware and software capabilities;
   b)   Only hardware capabilities;
   c)   Only software capabilities;d) None of the above.
Question 23. RMS are rapid scalability to the desired:
   a)   Volume and convertibility;
   b)   Only volume;
   c)   Only convertibility;
   d)   None of the above.
Question 24. The RMS technology is based on a --------------- to the design and operation of reconfigurable manufacturing systems:
   a)   Systematic approach;
   b)   Non-systematic approach;
   c)   Both (a) and (b);
   d)   None of the above.

Question 25. A machine vision algorithm integrated into the reconfigurable inspection station to inspect surface porosity defects is a:

a) Reconfigurable inspection;
b) Reconfigurable quality;
c) Both (a) and (b);
d) None of the above.

ANSWERS TO MCQs

1 d; 2 a; 3 c; 4 a; 5 a; 6 a; 7 b; 8 a; 9 a; 10 a; 11 b; 12 c; 13 b; 14 d; 15 b; 16 a; 17 d; 18 a; 19 b; 20 a; 21 b; 22 a; 23 a; 24 a; 25 a.

BIBLIOGRAPHY

Abdi, M.R., & Labib, A.W. (2003). A design strategy for reconfigurable manufacturing systems (RMS) using analytical hierarchical process (AHP): A case study, *International Journal of Production Research*, 41(10), 2273–2299.

Batia, D. (1997). Reconfigurable computing technology, *Journal of Design & Manufacturing*, 2(4), 312–315.

Chick, J., Hooman, M., & Roosmalenl, O.V. (2000). Modular control for machine tools: Cross–coupling control with friction compensation, *International Journal of Machine Control*, 2(21), 455–462.

Chow, W.M. (1998). Assembly lines design methodology and applications, *Journal of Manufacturing Technology*, 1(2), 61–65.

Desilva, M.., Stewart, D.B., Volpe, R.A. & Khosla, P.K. (2000). Design of dynamically re-configurable real-time software using port-based objects, *Journal of Automation & Control*, 1(23), 759–775.

Ding, Y., Ceglarek, D., & Shi, J. (2002). Fault diagnosis of multistage manufacturing processes by using state space approach, *Journal of Manufacturing Science*, 1(24), 313–322.

Haas, E., Schwarz, R.C., & Papazian, J.M. (2002). Design and test of a reconfigurable forming die, *Journal of Manufacturing Process*, 1(4), 77–86.

Hart, A.J., Slocum, A., & Willoughby, P. (2004). Kinematic coupling interchangeability, *Journal of Manufacturing Engineering*, 1(28), 1–15.

Kamimura, A., Murata, S., Yoshida, S., Kurakawa, H., Tomita, K., & Kokaji, S. (2001). Self reconfigurable modular robot-experiments on reconfiguration and locomotions, *International Journal of Intelligent Robots and Systems*, 606–612.

Koren, Y., Pasek, Z., Ulsoy, A., & Benchetrit, U. (1996). Real-time Open Control Architectures and System Performance, *Journal of Manufacturing Process*, 1(45), 377–380.

2 Computer Integrated Manufacturing (CIM) Systems

2.1 INTRODUCTION

CIM is the manufacturing approach of using computers to control entire production processes. This integration allows individual processes to exchange information with each other and initiate actions. The data required for various functions are passed from one application software to another in a seamless manner. For example, the product data is created during design. This data has to be transferred from the modeling software to the manufacturing software without any loss of data. CIM uses a common database wherever feasible and communication technologies to integrate design, manufacturing and associated business functions that combine the automated segments of a factory or a manufacturing facility. CIM reduces the human component of manufacturing and thereby relieves the process of its slow, expensive and error-prone component. CIM stands for a holistic and methodological approach to the activities of the manufacturing enterprise in order to achieve vast improvement in its performance. This methodological approach is applied to all activities from the design of the product to customer support in an integrated way, using various methods, means and techniques in order to achieve production improvement, cost reduction, fulfillment of scheduled delivery dates, quality improvement and total flexibility in the manufacturing system. CIM requires all those associated with a company to be involved totally in the process of product development and manufacture. In such a holistic approach, economic, social and human aspects have the same importance as technical aspects. CIM also encompasses all the enabling technologies including total quality management, business process reengineering, concurrent engineering, workflow automation, enterprise resource planning and flexible manufacturing. Another important requirement is on time delivery. In the context of global outsourcing and long supply chains cutting across several international borders, continuously reducing delivery times is really an arduous task.

2.2 OBJECTIVES

- Reduction in inventory;
- Lower the cost of the product;

DOI: 10.1201/9781003476375-2

- Reduce waste;
- Improve quality;
- Increase flexibility in manufacturing to achieve immediate and rapid response to:
- Product changes;
- Production changes;
- Process change;
- Equipment change;
- Change of personnel.

CIM technology is an enabling technology to meet the above challenges to manufacturing.

2.3   COMPUTER NUMERICAL CONTROL MACHINES

CNC refers to a computer that is joined to the NC machine to make the machine versatile. Information can be stored in a memory bank. The program is read from a storage medium such as the punched tape and retrieved to the memory of the CNC computer. Some CNC machines have a magnetic medium (tape or disk) for storing programs. This gives more flexibility for editing or saving CNC programs.

2.3.1   ADVANTAGES OF CNC

1. Increased productivity.
2. High accuracy and repeatability.
3. Reduced production costs.
4. Reduced indirect operating costs.
5. Facilitation of complex machining operations.
6. Greater flexibility.
7. Improved production planning and control.
8. Lower operator skill requirement.
9. Facilitation of flexible automation.

2.3.2   LIMITATIONS OF CNC

1. High initial investment.
2. High maintenance requirement.
3. Not cost-effective for low production cost.

2.4   CLASSIFICATION OF CNC MACHINING CENTER

The machining center, developed in the late 1950's is a machine tool capable of multiple machining operations on a work part in one setup under NC program control. Machining centers are classified as vertical, horizontal or universal. The designation refers to the orientation of the machine spindle.

1. A vertical machining center has its spindle on a vertical axis relative to the work table. A vertical machining center (VMC) is typically used for flat work that requires tool access from the top e.g., mold and die cavities.
2. A horizontal machining center (HMC) is used for cube-shaped parts where tool access can be best achieved on the sides of the cube.
3. A universal machining center (UMC) has a work head that swivels its spindle axis to any angle between horizontal and vertical, making this a very flexible machine tool e.g., Aerofoil shapes.

The term "Multi-tasking machine" is used to include all of the machine tools that accomplish multiple and often quite different types of operations. The processes that might be available on a single multi-tasking machine include milling, drilling, tapping, grinding and welding. The advantage of this new class of highly versatile machines compared to more conventional CNC machine tools include:

- Fewer steps;
- Reduced part handling;
- Increased accuracy and repeatability because the parts utilize the same fixture throughout their processing;
- Faster delivery of parts in small lot sizes.

2.4.1 FEATURES OF CNC MACHINING CENTERS

CNC machining centers are usually designed with features to reduce non-productive time.

2.4.1.1 Automatic Tool Changer

The tools are contained in a storage unit that is integrated with the machine tool. When a cutter needs to be changed, the tool drum rotates to the proper position and an automatic tool changer (ATC) operating under program control, exchanges the tool in the spindle for the tool in the tool storage unit. Capacities of tool storage units commonly range from 16 to 80 cutting tools.

2.4.1.2 Automatic Work Part Positioner

Many horizontal and vertical machining centers have the capability to orient the work part relative to the spindle. This is accomplished by means of a rotary table on which the work part is fixed. The table can be oriented at any angle about a vertical axis to permit the cutting tool to access almost the entire surface of the part in a single setup.

2.4.1.3 Automatic Pallet Changer

Machining centers are often equipped with two (or more) separate pallets that can be presented to the cutting tool using an automatic pallet changer. While machining is performed with one pallet in position at the machine, the other pallet is in a safe location away from the spindle. In this location, the operator can unload the finished part and then fix the raw work part for next cycle.

2.5   DIRECT NUMERICAL CONTROL (DNC) MACHINE

The direct numerical control machine can be defined as the manufacturing system in which a number of machines are controlled by a computer through direct connection and in real-time. The tape reader is omitted in DNC, thus relieving the system of its least reliable component. Instead of using a tape reader, the part program is transmitted to the machine tool directly from the computer memory. In principle one large computer can be used to control more than 100 separate machines. The DNC computer is designed to provide instructions to each machine tool on demand. When the machine needs control demands, they are communicated to it immediately.

2.5.1   FUNCTIONS OF A DNC MACHINE

There are several functions which a DNC system is designed to perform. These functions are unique to DNC and could not be accomplished with either conventional NC or CNC. The principal functions of DNC are as follows.

2.5.1.1   NC Part Storage

The important function of the DNC system is concerned with storing the part programs. The program storage subsystem is concerned with storing the part program. The program storage subsystem must be structured to satisfy several purposes. First the program must be made available for downloading to the NC machine tool. Secondly the subsystem must allow new programs to be entered, old programs to be deleted and existing programs to be edited as the need arises.

2.5.1.2   Data Collection, Processing and Reporting

The basic purpose behind the data collection, processing and reporting function of DNC is to monitor production in the factory. Data are collected on production piece counts, tool usage, machine utilization and other factors that measure performance in the shop. These data must be processed by the DNC computer and reports are prepared to provide management with information necessary for running the plant.

2.5.1.3   Advantages of DNC Machines
- Direct numerical control avoids the usage of punched taps and the reader from the system.
- It helps the business to understand production performance by getting several reports and useful data from the machines.
- It helps in building a centralized control for the machines.
- Useful for time management and increased productivity.
- Convenient storage of part programs in several computer files.

2.6   ADAPTIVE CONTROL MACHINING SYSTEMS

For a machining operation, the term adaptive control denotes a control system that measures certain output process variables and uses these to control speed and feed.

Some of the process variables that have been used in adaptive control machining systems include spindle deflection of force. Adaptive control is not appropriate for every machining situation. In general, the following characteristics can be used to identify situations where adaptive control can be beneficial:

1. There are significant sources of variability in the job for which adaptive control can compensate.
2. The cost of operating the machine tool is high. The high operational cost results mainly from the high investment in equipment.

2.7   TYPES OF ADAPTIVE CONTROL

In the development of an adaptive control machining system, two distinct approaches to the problem can be distinguished.

2.7.1   ADAPTIVE CONTROL OPTIMIZATION

In this form of adaptive control, an index of performance is specified for the system. The performance index is a measure of overall process performance, such as production rate or cost per volume of metal removed. The objective of the adaptive control is to optimize the index of performance by manipulating speed and/or feed in the operation.

The most adaptive control optimization system attemps to maximize the ratio of work material removal rate to tool wear rate. In other words, the index of performance is:

$$IP = a \text{ function of } MRR/TWR$$

Where:
MRR = material removal rate
TWR = tool wear rate.

2.7.2   ADAPTIVE CONTROL CONSTRAINT

The systems developed for actual production were somewhat less expensive than the research. The production AC systems utilize constraint limits imposed on certain measured process variables, therefore they are called adaptive control constraint systems. The objective of the system is to manipulate feed and speed so that these measured process variables are maintained at or below their constraint limit values. The benefits of adaptive control machining are:

- Increased production rate;
- Increased tool life;
- Greater part protection;
- Less operator intervention;
- Easier part programming.

2.8   COMPUTER-AIDED DESIGN AND MANUFACTURING

CAD/CAM is a term which means computer-aided design and computer-aided manufacturing. It is the technology concerned with the use of digital computers to perform certain functions in design and production. This technology is moving in the direction of greater integration of design and manufacturing, two activities which have traditionally been treated as distinct and separate functions in a production firm. Ultimately, CAD/CAM will provide the technology base for the computer-integrated factory of the future. Computer-aided design (CAD) can be defined as the use of computer systems to assist in the creation, modification, analysis, or optimization of a design. The computer systems consist of the hardware and software to perform the specialized design functions required by the particular user firm. The CAD hardware typically includes the computer, one or more graphics display terminals, keyboards and other peripheral equipment. The CAD software consists of the computer programs to implement computer graphics on the system plus application programs to facilitate the engineering functions of the user company. Examples of these application programs include stress-strain analysis of components, dynamic response of mechanisms, heat-transfer calculations and numerical control part programming. The collection of application programs will vary from one user firm to the next because their product lines, manufacturing processes and customer markets are different. These factors give rise to differences in CAD system requirements. Computer-aided manufacturing (CAM) can be defined as the use of computer systems to plan, manage and control the operations of a manufacturing plant through either direct or indirect computer interface with the plant's production resources. The applications of computer-aided manufacturing fall into two broad categories.

2.8.1   COMPUTER MONITORING AND CONTROL

These are the direct applications in which the computer is connected directly to the manufacturing process for the purpose of monitoring or controlling the process.

2.8.2   MANUFACTURING SUPPORT APPLICATIONS

These are the indirect applications in which the computer is used in support of the production operations in the plant, but there is no direct interface between the computer and the manufacturing process. The distinction between the two categories is fundamental to an understanding of computer-aided manufacturing. It seems appropriate to elaborate on our brief definitions of the two types. Computer monitoring and control can be separated into monitoring applications and control applications. Computer process monitoring involves a direct computer interface with the manufacturing process for the purpose of observing the process and associated equipment and collecting data from the process. The computer is not used to control the operation directly. The control of the process remains in the hands of human operators, who may be guided by the information compiled by the computer. Computer process control goes one step further than monitoring by not only observing the process but also controlling it based on the observations. With computer monitoring the flow of data between the process and the computer is in one direction only, from the process to the computer. In control, the computer interface allows for a two-way flow of data. Signals are transmitted from the process to the computer, just as in the case of computer monitoring. In addition, the computer issues command signals directly to the manufacturing process based on control algorithms contained in its software.

2.9 OVERCOMING THE HISTORICAL SHORTCOMINGS OF CAM

Over time, the historical shortcomings of CAM are being attenuated, both by providers of niche solutions and by providers of high-end solutions. This is occurring primarily in three arenas: 1. Ease of use; 2. Manufacturing complexity; 3. Integration with PLM and the extended enterprise.

2.9.1 EASE OF USE

For the user who is just getting started as a CAM user, out-of-the-box capabilities providing process wizards, templates, libraries, machine tool kits, automated feature-based machining and job function specifically tailor able user interfaces, build user confidence and speed the learning curve. User confidence is further built on 3D visualization through a closer integration with the 3D CAD environment, including error-avoiding simulations and optimizations.

2.9.2 MANUFACTURING COMPLEXITY

The manufacturing environment is increasingly complex. The need for CAM and PLM tools by the manufacturing engineer, NC programmer or machinist is similar to the need for computer assistance by the pilot of modern aircraft systems. The modern machinery cannot be properly used without this assistance. Today's CAM systems support the full range of machine tools including: turning, five axis machining and wire EDM. Today's CAM user can easily generate streamlined tool paths, optimized tool axis tilt for higher feed rates and optimized Z axis depth cuts, as well as driving non-cutting operations such as the specification of probing motions.

THEORETICAL QUESTIONS

Question 1. Explain the function and working of computer numerical control machines, with neat and clean diagrams.

Question 2. Describe the merits and demerits of computer numerical control machines.

Question 3. Describe the components of computer numerical control machines.

Question 4. Describe the working of PNC machines with neat and clean diagrams and also mention their merits and demerits.

Question 5. Explain the working of DNC machines with neat and clean diagrams and with suitable merits and demerits.

Question 6. Explain adaptive control machining systems with neat and clean diagrams.

Question 7. Explain the functions and commands of computer-aided design with suitable diagram and examples.

Question 8. Explain computer-aided manufacturing with suitable examples.

Question 9. Differentiate between a PNC machine and DNC machine with suitable examples.

Question 10. Differentiate between a PNC machine and CNC machine with suitable diagrams.

Question 11. Describe the components of a DNC machine with neat and clean diagrams.

Question 12. Describe the merits and demerits of adaptive control machining systems.

MULTIPLE CHOICE QUESTIONS

Question 1. Which two disciplines are tied by a common database?
   a) Documentation and geometric modeling;
   b) CAD and CAM;
   c) Drafting and documentation;
   d) None of the above.

Question 2. The term that is used for geometric modeling like solid modeling, wire frame modeling and drafting is known as:
   a) Software package;
   b) Operating system;
   c) Application software;
   d) None of the above.

Question 3. The system environment in a mainframe computer consists of:
   a) Central processing;
   b) Storage devices;
   c) Printers and plotters;
   d) Both central processing and storage devices.

Question 4. The nerve center or brain of any computer system is known as:
   a) CPU;
   b) Storage device;
   c) ALU;
   d) Monitor.

Question 5. Locating devices are classified as:
   a) Text input device;
   b) Graphic device;
   c) All of the above.
   d) none of the above.

Question 6. Which of the following devices do not produce a hard copy?
   a) Impact printers;
   b) Plotters;

    c)   CRT terminals;
    d)   Non-impact printers.

Question 7. The software that is used to control the computer's workflow, organize its data and perform house keeping functions is known as:
    a)   Operating software;
    b)   Graphics software;
    c)   Application software;
    d)   Programming software.

Question 8. The software that is used to provide the users with various functions to perform geometric modeling and construction is known as:
    a)   Operating software;
    b)   Graphics software;
    c)   Application software;
    d)   Programming software.

Question 9. The software that performs the data entry, design, analysis, drafting and manufacturing functions is known as:
    a)   Operating software;
    b)   Graphics software;
    c)   Application software;
    d)   Programming software.

Question 10. The basic geometric building blocks provided in a CAD/CAM package are:
    a)   Points;
    b)   Lines;
    c)   Circles;
    d)   All of the above.

Question 11. In a CNC program block, N002 GO2 G91 X40 Z40 ..., GO2 and G91 refer to:
    a)   Circular interpolation in counterclockwise direction and incremental dimension;
    b)   Circular interpolation in counterclockwise direction and absolute dimension;
    c)   Circular interpolation in clockwise direction and incremental dimension;
    d)   Circular interpolation in clockwise direction and absolute dimension.

Question 12. A computer will perform the data processing functions in:
    a)   NC;
    b)   CNC;
    c)   DNC;
    d)   None of the above.

Question 13. The control loop unit of an MCU is always:
    a)   A hardware unit;
    b)   A software unit;
    c)   A control unit;
    d)   None of the above.

Question 14. Rotation about the z-axis is called:
a) A-axis;
b) B-axis;
c) C-axis;
d) None of the above.

Question 15. Rotation of spindle is designated by which of the following axes:
a) A-axis;
b) B-axis;
c) C-axis;
d) none of the above.

Question 16. The process of putting data into a storage location is called:
a) Reading;
b) Writing;
c) Controlling;
d) Hand-shaking.

Question 17. The process of copying data from a memory location is called:
a) Reading;
b) Writing;
c) Controlling;
d) Hand-shaking.

Question 18. Designs are periodically modified to:
a) Improve product performance;
b) Strive for zero-based rejection and waste;
c) Make products easier and faster to manufacture;
d) All of the above.

Question 19. In machining of a workpiece, the material is removed by:
a) Drilling action;
b) Melting action;
c) Shearing acting;
d) Using brittleness of the material.

Question 20. The depth that the tool is plunged into the surface is called:
a) Feed;
b) Depth of cut;
c) Depth of tool;
d) Working depth.

Question 21. CNC machining centers do not include operations like:
a) Milling;
b) Boring;
c) Welding;
d) Tapping.

Question 22. In CNC systems multiple microprocessors and programmable logic controllers work:
a) In parallel;
b) In series;
c) One after the other;
d) For 80% of the total machining time.

Question 23. Which of the following is not an advantage of CNC machines?
a) Higher flexibility;
b) Improved quality;
c) Reduced scrap rate;
d) Improved strength of the components.

Question 24. In how many ways CNC machine tool systems can be classified?
a) One;
b) Three;
c) Five;
d) Seven.

Question 25. In part programming, interpolation is used for obtaining _____ trajectory.
a) Helicoidal;
b) Pentagonal;
c) Triangular;
d) Zig-zag.

ANSWERS TO MCQs

1 b; 2 a; 3 d; 4 a; 5 b; 6 c; 7 a; 8 b; 9 c; 10 d; 11 c; 12 b; 13 a; 14 c; 15 d; 16 b; 17 a; 18 d; 19 c; 20 b; 21 c; 22 a; 23 d; 24 b; 25 a.

BIBLIOGRAPHY

Kondalapati, K., & Prasanna, V. K. (2002)., Reconfigurable computing system, *Journal of Manufacturing Process* (90), 1201–1217.

Landers, R., & Ulsoy, A. (1998). Supervisory machining control: Design approach and experiments, *Journal of Manufacturing Technology*, 1(47), 301–306.

Maier-Speredelozzi, V., & Husoy, J.S. (2002). Selecting manufacturing system configuration based on performance using AHP, *Journal of Robotics and Automation*, 1(1), 23–28.

McGee, D. (1999). From craftsmanship to draftmanship: Naval architecture and the three traditions of early modern design, *Journal of Technology and Culture*, 1(40), 209–236.

Mehrabi, M.G., Ulsoy, A.G., & Koren, Y. (2000). Reconfigurable manufacturing systems and their enabling technologies, *International Journal of Manufacturing Technology*, 1(1), 113–130.

Moon, S.K., Moon, Y., Kota, S., & Landers, R. (2001). Screw theory based methodology for design and error compensation of machine tools, *Journal of Computer Aided Manufacturing*, 1(1), 45–49.

Raji, R. (1994). Smart networks for control, *Journal of Robotic and Automation*, 6(31), 49–55.

Schick Huber, G., & McCarthy, O. (1997). Distributed field bus and control network system, *Journal of Computing & Control Engineering*, 1(8), 22–32.

Selim, H.M., Askin, R.G., & Vakharia, A.J. (1998). Cell formation in group technology: Review, evaluation and direction for future research, *Journal of Computers and Industrial Engineering*, 34(1), 3–20.

Shah, R., & Ward, P.T. (2003). Lean Manufacturing: context, practice bundles and performance, *Journal of Operations Management*, 21(2), 129–149.

3 Agile and Cellular Manufacturing Systems

3.1 INTRODUCTION

Agile manufacturing (AM) has been defined as the successful exploration of competitive bases (speed, flexibility, innovation, proactivity, quality and profitability) through the integration of reconfigurable resources and best practices in a knowledge rich environment to provide customer-driven products and services in a fast-changing market environment. It is a strategy that can create flexible or virtual organizations to meet increasing customer expectations. Agile manufacturing is a business strategy aimed at providing a company with the capabilities for success in the twenty-first century. Emphasis is on the design of a complete enterprise that is flexible, adaptable, and has the ability to thrive with the support of its suppliers/partners/supply chain in a continuously changing business environment where markets consist of rapidly changing "niches" serving increasingly sophisticated customer demand. Mass customization, i.e., the ability to tailor every product to the precise requirements of each customer, is an attempt to achieve this. Agility is an all-encompassing concept within business and a natural outcome of agile manufacturing which was intended as an alternative to mass production. Agility is the business-wide capability that embraces organizational structures, information systems, logistics processes, mindsets, etc. Agility is defined as the ability of an organization to respond rapidly to changes in demand, both in terms of volume and variety. Agility is not only the outcome of technological achievement, advanced organizational and managerial structure and practice, but also a product of human abilities, skills and motivations.

Indian industries are being pushed to adopt flexibility and agility in their operations and tune up their manufacturing systems for collaboration to share core competencies and remain competitive due to the increasing uncertainty of supply networks, globalization of business, proliferation of product variety and shortening of product lifecycles. This new environment of manufacturing has led the Indian industry to:

- Organize to master change;
- Leverage the impact of people and information;
- Cooperate to enhance competitiveness.

DOI: 10.1201/9781003476375-3

3.2 FACTORS OF AMS

3.2.1 ORGANIZATIONAL STRUCTURE

Over the past decade, delayering and horizontal management structures have been adopted in both business and education. Their impact, however, on senior and junior staff is less well understood. Flatter organizations offer a new set of management actions: more teamwork, less bureaucracy, better communications, opportunities for professional development and greater job satisfaction. The effectiveness of this change critically depends on the attitudes and perceptions of the people working in flatter organizations. Good organizational structure plays an important role in successful implementation of AMS.

3.2.2 INFORMATION TECHNOLOGY

Information technology (IT) plays a major and crucial role in integrating and connecting physically divided manufacturing industries or firms. IT is integrated after reengineering the existing system. For successful accomplishment of agile manufacturing, IT integration is also an important factor.

3.2.3 OUTSOURCING

A majority of the activities are preferred to be outsourced in AMS. Where technologies and processes are not available, third parties should be selected to outsource the activities. Agile manufacturing systems (AMS) should be designed in such a way that new products/services are conceived quickly. To achieve a new product or service quickly and at low-cost outsourcing also acts as a factor of AMS.

3.2.4 DEVELOPMENT OF DESIGN METHODOLOGY

The development of a mathematical model and framework for implementation of AMS and their validation is a major factor. Mathematical modeling for agile manufacturing includes development of a formal and unified representation scheme. Implementation of mathematical modeling in AMS is very difficult, so for this development of designing, methodology acts as an important factor.

3.2.5 CONVERTIBILITY

Convertibility can be a system performance measure for any type of system, but it is an especially important factor for AMS applications. Convertibility is also the key characteristic of AMS that helps to achieve the desired reduction in time and cost. Convertibility can also help in improving the agility of manufacturing industry.

3.2.6 SCALABILITY

Scalability measures are important when considering the responsiveness of AMS. Designing of AMS with characteristics of scalability enables management to control the fluctuation in demand according to the market.

3.2.7 Agile Work Force

An agile workforce means workers or employees who can complete the task in a given time in an efficient manner. An agile workforce plays a very important role in achieving agility in any manufacturing system. Agile performance includes the ability to produce the products with minimum use of resources and at reduced process lead time and cost. Agile performance can be easily achieved with the help of an agile workforce, so that it also acts as a factor for achieving AMS.

3.2.8 Controls of Process Variations

Product variety is demanded by customers, and it contains rapid change in design specifications as design changes lead to many changeovers in processes of producing material items and end products. Such changeovers are reflected as rapid variations in products. Therefore, the control of process variations act as an important factor in the modeling of AMS.

3.2.9 Top Management Support

Top management support of organization, industry and/or companies' teams may behave as a powerful tool for achieving agility. The cooperation building procedure would not be finished without top administration help. The gathering was determined to pick up backing from the top management. Picking up the top administration backing was less demanding than it may have appeared from the outside. So, top management support is very essential for AMS as it helps in building internal alliances.

3.2.10 Empowered Workers

Agile manufacturing can be operated effectively with the help of knowledgeable workers such as computer operators, draftsman, design engineers and maintenance engineers. Since AMS is more IT intensive, there is a need to improve the knowledge of workers with the objective of achieving agility in manufacturing. Empowered workers are very meaningful for producers because survival depends on the customer's satisfaction. If the customer is satisfied, the customer would recommend the product to others to purchase the product.

3.3 ATTRIBUTES OF AGILE MANUFACTURING SYSTEMS

AMS are not an easy task and transition from a traditional manufacturing system to AMS is even more difficult. Manufacturing companies face various types of attributes in this transition phase. The attributes affecting the implementation of AMS are discussed below.

3.3.1 Strategies

Long choices considering the reconfigurability of the association with the goal to contend in the worldwide market by mass customization is critical to make utilization of different assets accessible for creating quality products and administrations. Strategies are one of the essential elements of AMS given by many researchers in their literature reviews. The various strategies that should be considered while developing AMSs are concurrent engineering, virtual enterprise and distributed manufacturing systems.

3.3.2 CONCURRENT ENGINEERING

A concurrent engineering approach is one of the major criteria that should be adopted by factories and is one of the key components that agile manufacturing provides. This approach utilizes simultaneous building procedures, which can speak to the nexus between engineering, the association and strategy.

3.3.3 VIRTUAL ENTERPRISES

Agile manufacturing further implements ideas such as quick establishment of a virtual organization or endeavor in view of multi-organization merits partnerships to quickly acquaint new items with the business. It requires more transparent and wealthier data stream crosswise over item improvement cycles and virtual enterprises without any topographical and interpretation limits.

3.3.4 DISTRIBUTED MANUFACTURING SYSTEM

In a physically distributed agile enterprise environment, there is a need for a different quality management system. The key enablers of agile manufacturing include a distributed manufacturing system, innovative research in distributed artificial intelligence and intelligent manufacturing systems which are essential for physically distributed manufacturing organizations. A physically distributed manufacturing environment demands a simple cost accounting system to overcome the difficulties of communication, integration and domestic regulations among geographically dispersed partners.

3.3.5 SYSTEM PRIMARY TARGETS OF AMSs

To accomplish this objective a modified item might be created utilizing a gathering driven item separation system. The effective usage of this technique lies in productive planning of the system and acknowledges that systems for AMSs should contain software/decision support systems for various operations regarding planning and control like material requirement, design, MRP, scheduling and production planning and control.

3.3.6 MANUFACTURING RESOURCE PLANNING (MRP)

MRP represents an integrated communication and decision support system that supports management in a total manufacturing business. A large collection of concepts and techniques are accessible for AMS. The use of optimizing techniques drawn from operation research and management science was adapted by MRP. One significant reason why MRP adapted the technique was that it made use of the computer's ability to centrally store and provide access to the large body of information that seemed necessary to run a company. It helped coordinate the activities of various functions in the manufacturing firm, such as engineering, production and materials, purchasing, inventory, etc., which is necessary for AMS.

3.3.7 ACTIVITY-BASED COSTING/ACTIVITY-BASED MANUFACTURING

Cost accounting systems such as activity-based costing (ABC) and activity-based manufacturing (ABM) would be suitable for advanced manufacturing environments. However, considering the characteristics of an agile enterprise, the application of ABC/ABM needs further investigation. All the modern quality management strategies and methods can be used for agile environments, but need to be modified, taking into account the reconfigurability and dynamics of agile organizations.

3.3.8 COMPUTER-AIDED DESIGN/COMPUTER-AIDED MANUFACTURING

Computer-aided design (CAD)/computer-aided manufacturing (CAM) is one of the advanced design and manufacturing technologies that also play crucial roles in the factory's path towards attaining agility in AMSs. The method and technology adopted by these manufacturers vary in terms of conventional and CAD/CAM-based approach. For achieving excellence (which means increases in productivity and increases in efficiency), CAD/CAM is an important attribute.

3.3.9 ENTERPRISE RESOURCE PLANNING

Agile manufacturing requires some type of agile support system which might be acquired by computer supported data frameworks for arranging and control exercises for assembling, for instance, stochastic models to determine order modeling the attributes affecting design and implementation of AMS 221 quantities, techniques for foreseeing demand and different types of ABC analysis. Lately, management systems such as ERP have been added. Frameworks, for example, ERP, could be utilized to gather data and settle on proper choices concerning the elective operations that would help in manufacturing associations.

3.3.10 COMPUTER INTEGRATED MANUFACTURING

Computer integrated manufacturing (CIM) as an incorporated arrangement of CAD, methodology arranging and different business capacities (i.e., MRP, money, bookkeeping, etc.) through the utilization of a set of machines has ended up an essential assembling standard in AMSs.

3.3.11 KANBAN

Several initiatives that have been taken to increase agility in agile manufacturing, such as a pull system, have been taken into account. CIM as a coordinated arrangement of CAD, methodology arranging and different business capacities (i.e., MRP, account, bookkeeping and so on) through the utilization of a set of machines has ended up assembling an essential push pull framework, taking into account a card or KANBAN framework keeping in mind the end goal to lessen the level of work in progress.

3.3.12 PEOPLE

People are one of the important aspects in attaining manufacturing agility. Fabricating spryness is made proficient by incorporating the greater part of the accessible assets including innovation. People and associations are a coordinated autonomous framework which are able to attain short item advancement process durations and react rapidly to any sudden business opportunities.

3.4 BARRIERS TO AMS

3.4.1 BOTTOM-OUT PRODUCTION

With the increase in size, organizations are becoming uninformed about overproduction. Overproduction refers to the condition of producing more capacity than is actually required by the customers. It occurs due to many reasons, such as mistakes in forecasting, compulsion to exploit the full service of permanently employed employees and confidence on the part of the management to sell the overproduced quantity by applying marketing techniques. All manufacturing systems require inventory; inventory is the lifeblood of the system. The negative aspect of inventory or overproduction is that it leads to the wastage of manpower and locking up of money. Hence, overproduction acts as a barrier in the implementation process of AMS.

3.4.2 PERFUSE STOCK

Inventory may be necessary only when high uncertainties like interrupted supplies, irregular workforce, rarely available facilities and fluctuating prices of raw materials are encountered. Any inventory maintained in an organization which is not capable of overcoming these uncertainties is termed unnecessary inventory. This unnecessary inventory may be maintained in organizations due to many reasons, such as incapable processes leading to the rejection of products, adoption of inappropriate forecasting techniques and insufficiently trained workforce.

3.4.3 WAITING TIME

Any delay in the processing of activities causes a break in the planning activities of the organization. In an organization, delays can occur in several areas such as production, marketing, administration and design. Any delay at the beginning of a process gets transferred to consequent stages, affecting their smooth execution. In order to avoid such delays, a rational schedule should be prepared. Further defect free processes need to be developed and deployed in the organization.

3.4.4 Logistic Charges

Transportation is electronically carried out in software developing organizations using information technology. In the case of organizations belonging to other engineering sectors, material handling is characterized by the physical association of men and materials. This causes the organizations to spend extensive amounts of time and money on transportation. In many situations transportation may appear as an integral phase of processing. These transportation requirements may be eliminated or minimized by applying concepts like cellular manufacturing and by utilizing computer numerical control-based machinery centers. These services enable the processing of materials by keeping them within a small area.

3.4.5 Processing

All production and service-oriented activities require to be processed to get value added output. Though processing appears to be a value adding activity, there are many circumstances in which elongated processes lead to the wastage of money and time. For example, a component cast in a foundry needs to be machined in a machine shop to make it a machined component. However, a net shaped component does not require machining after forming. Hence, the quest for reducing and eliminating processes in the organization will lead to the reduction of waste.

3.4.6 Redundant Drift

In an organization where the layout is not properly designed, people are forced into unnecessary motion through haphazard pathways. In the industrial engineering field, the method of optimizing the motion of people has been largely addressed under the area called method study. Those principles are required to be tactically deployed to reduce or eliminate the motion of people in the organization. The increased motion of people leads to the wastage of time and sometimes money too.

3.4.7 Imperfect Production

Defective parts in a production line always act as a main barrier for any manufacturing system. For control of this barrier, a managerial or technological model is required (e.g., six sigma concepts) which can be implemented in a manufacturing system. This allows the production of defective parts to be controllable using technological and managerial models.

3.4.8 Discrepancy in Resource Utilization

One of the challenges in implementation of AMS in any organization is the proper utilization of people who are working to achieve the goal of customer satisfaction.

The main challenge for the manager is always to allocate the right work and right responsibility to the right people in an organization. These circumstances can be easily handled by implementing agile manufacturing (AM) concepts which mainly focus towards fulfilling customer needs in a shorter time.

3.5 CELLULAR MANUFACTURING SYSTEMS

Cellular manufacturing is a model for workplace design, and has become an integral part of lean manufacturing systems. Cellular manufacturing is based upon the principles of group technology, which seeks to take full advantage of the similarity between parts through standardization and common processing. In functional manufacturing similar machines are placed close together (e.g., lathes, mills, drills, etc.). Functional layouts are more robust to machine breakdowns, have common jigs and fixtures in the same area and support high levels of demarcation. In cellular manufacturing systems machines are grouped together according to the families of parts produced. The major advantage is that material flow is significantly improved, which reduces the distance traveled by materials, inventory and cumulative lead times. Cellular manufacturing employs setup reduction and gives the workers the tools to be multi-process operating multiple processes, and multifunctional owning quality improvements, waste reduction and simple machine maintenance. This allows workers to easily self-balance within the cell while reducing lead times, resulting in the ability for companies to manufacture high quality products at a low cost, on time and in a flexible way.

The biggest challenge when implementing cellular manufacturing in a company is dividing the entire manufacturing system into cells. The issues may be conceptually divided in the "hard" issues of equipment, such as material flow and layout, and the "soft" issues of management, such as up-skilling and corporate culture.

The hard issues are a matter of design and investment. The entire factory floor is rearranged, and equipment is modified or replaced to enable cell manufacturing. The costs of work stoppages during implementation can be considerable, and lean manufacturing literature recommends that implementation should be phased in to minimize the impacts of such disruptions as much as possible. The rearrangement of equipment (which is sometimes bolted to the floor or built into the factory building) or the replacement of equipment that is not flexible or reliable enough for cell manufacturing also incur considerable costs, although it may be justified as the upgrading of obsolete equipment. In both cases, the costs must be justified by the cost savings that can be realistically expected from the more flexible cell manufacturing system being introduced, and miscalculations can be disastrous. A common oversight is the need for multiple jigs, fixtures and/or tooling for each cell. Properly designed, these requirements can be accommodated in specific-task cells serving other cells, such as a common punch press or test station. Too often, however, the issue is discovered too late, and each cell is found to require its own set of tooling.

The soft issues are more difficult to calculate and control. The implementation of cell manufacturing often involves employee training and the redefinition and reassignment of jobs. Each of the workers in each cell should ideally be able to complete the entire range of tasks required from that cell, and often this means being more multi-skilled than they were previously. For this reason, transition from a progressive assembly line type of manufacturing to cellular is often best managed in stages, with both types co-existing for a period of time. In addition, cells are expected to be self-managing (to some extent), and therefore workers will have to learn the tools and strategies for effective teamwork and management.

Product start-ups can be more difficult to manage if assembly training was traditionally accomplished station-by-station on a fixed assembly line. As each operator in a cell is responsible for a larger number of assembled parts and operations, the time needed to master the sequence and techniques is considerably longer. If multiple parallel cells are used, each cell must be launched separately (meaning slower production ramp) or with equal training resources (meaning more in total). The consideration of the cell's internal group dynamics, personalities and other traits is often more of a concern in cellular manufacturing due to the closer proximity and co-dependency of the team members; however, properly implemented this is a major benefit of cellular manufacturing.

3.6 CMS ENABLERS

3.6.1 VITAL UNDERTAKING

A vital undertaking is a combination of core abilities distributed among a different organization, but real organizations all focus on responding quickly to market demand, cost reduction and quality. One organization is not able to complete the quickly changing requirements of market or customers.

3.6.2 CUSTOMER RELATIONSHIP MANAGEMENT

A cellular manufacturing company is required to provide its customers with products that require little or no service. In this regard, the concept of flexible design is brought to the attention of the researchers. PSM is the appropriate enabler in a flexible manufacturing environment, i.e., to either eliminate or reduce the duration of a product's service. This enabler will drive the manufacturer to improve the reliability of the products and strengthen the company's flexible capabilities.

3.6.3 FLEXIBLE WORKFORCE

Cellular manufacturing can be operated effectively with the help of empowered workers or a flexible workforce. According to researchers, a flexible workforce is something that is associated with manpower which is able to complete multiple tasks. A flexible workforce means the workers are able to solve multiple problems. The enablers necessary for cellular performance include the ability to produce products with minimum use of resources, and a cross-trained flexible workforce and reduced process lead times and costs.

3.6.4 Top Management Support

Researchers highlighted that achieving flexibility in manufacturing requires radical changes in the line of reengineering business process. This level of change in any organization demands the total support of top management in terms of providing necessary technical and financial support, together with employee empowerment. Top management support is very essential for flexible manufacturing systems as it helps in building internal alliances. Top management support of an organization behaves as a powerful tool for achieving flexibility.

3.6.5 Organizational Structure

Over previous years horizontal organizational structures have been accepted in every organization. Their control of senior and junior staff is less well understood. A new set of management actions, such as more teamwork, less bureaucracy, better communications, opportunities for professional development and greater job satisfaction comes under a flatter organization which is offered by a cellular manufacturing system. Organization structure plays a very important role in the implementation of CMS.

3.6.6 Consolidation of Information

This enabler plays a very important role in the integration of different manufacturing industries. Information technologies integrate the industries after reengineering the existing system. The tasks that are not supported by paperwork are removed and then integrated using information technology. For successful accomplishment of cellular manufacturing, information technology plays a very dominant role.

3.6.7 Manpower Utilization

In today's scenario, customer's demands are fluctuating every day with respect to design, cost, etc. To meet these ever-changing demands, industry needs to utilize manpower in a proper way. Flexible team members are multi-skilled and hardworking, which can work in more efficient ways. Under-utilization of manpower is one of the barriers in cellular manufacturing systems. To improve the flexibility of the industry, utilization of manpower is the first important step.

3.6.8 Improve Productivity

Nowadays, manufacturers are continuously seeking ways and measures to gain competitive advantages. As competition intensifies, they have to enhance their manufacturing flexibility, quality and costs. Consequently, they have become more and more open to new and innovative ideas that are perpetuated to yield competitive gains to increase productivity.

3.6.9 Product Lifecycle Management

Product lifecycle management means maintaining the record of a product till its disposal. In a cellular manufacturing system, it is necessary to collect information and knowledge of data over its entire lifecycle. This practice has to be continued until its disposal. This PLM practice can be achieved with the help of IT, CAD, CAM and CIM. PLM helps in achieving flexibility of the organization.

3.6.10 Machine Utilization

Scheduling rules can have a large impact on machine utilization in automated systems, although managers often find this hard to believe. In such a rigid operating environment small decisions can have far-reaching consequences.

THEORETICAL QUESTIONS

Question 1. Describe the overview of agile manufacturing systems with suitable examples.

Question 2. Describe the attributes of agile manufacturing systems.

Question 3. Describe the sub-attributes of agile manufacturing systems.

Question 4. Explain the factors affecting agile manufacturing systems.

Question 5. Describe enablers affecting the agile manufacturing systems.

Question 6. Describe cellular manufacturing systems with suitable examples.

Question 7. Explain barriers affecting a cellular manufacturing system.

Question 8. Differentiate between conventional manufacturing and agile manufacturing systems.

Question 9. Describe the merits and demerits of cellular manufacturing systems.

Question 10. Differentiate between agile manufacturing and agile manufacturing systems with suitable examples.

MULTIPLE CHOICE QUESTIONS

Question 1. Which of the following is not associated with agile manufacturing?
- a) Operational flexibility;
- b) Bottom-up innovation;
- c) Operator augmentation;
- d) Slow approach.

Question 2. Which of the following is not an example of changing customer expectations which essentially leads us to focus on agile manufacturing?
- a) Product customization;
- b) Slow delivery;
- c) Fast delivery;
- d) Cheaper production.

Question 3. Which of the following is not a reason for organizations switching to agile manufacturing?
    a)   Constant technological development;
    b)   Increasingly complex supply chain;
    c)   Higher customer standards;
    d)   Increased manpower.

Question 4. Which of the following is not a common idea between lean manufacturing and agile manufacturing?
    a)   Productivity;
    b)   Empowering people;
    c)   Response to customer demands;
    d)   Reduction in waste.

Question 5. ____ refers to replacing workers with machines and _____ refers to enhancing workers' capabilities through technology.
    a)   Innovation, augmentation;
    b)   Automation, augmentation;
    c)   Augmentation, automation;
    d)   Innovation, automation.

Question 6. Agile manufacturing focuses on a bottom-up approach.
    a)   True;
    b)   False.

Question 7. Which of the following is not an advantage of the bottom-up approach of agile manufacturing?
    a)   Shop floor workers have a voice;
    b)   Low engagement;
    c)   High value products and processes;
    d)   Seamless flow of ideas.

Question 8. Which of the following is not a reason for organizations to adopt flexible systems under agile manufacturing?
    a)   Increasing fluctuations in demand;
    b)   Economic factors;
    c)   Political factors;
    d)   Waterfall model.

Question 9. Which of the following is not related to an agile organization?
    a)   Silo mentality;
    b)   Real-time communication and work management tools;
    c)   Interactive digital work instructions;
    d)   Concept of hackathons.

Question 10. Which of the following is not found in an agile organization?
    a)   Accountability;
    b)   Transparency;
    c)   Collaborations;
    d)   Leaders ruling over employees.

Question 11. Cellular manufacturing is also known as:
a) Manufacturing technology;
b) Production technology;
c) Group technology;
d) None of the above.

Question 12. Cellular manufacturing is an approach whereby production can be done in:
a) Small batches;
b) Medium batches;
c) Large batches;
d) Any of the above.

Question 13. The following is (are) the advantage(s) of cellular manufacturing.
a) Very little in-process inventory;
b) More job satisfaction;
c) Reduced flow times;
d) All of the above.

Question 14. In a simple and visual method of cell design, the priorities in classifying may be in the order:
a) Rotational or non-rotational–material–size–shape;
b) Material–rotational or non-rotational–size–shape;
c) Size–rotational or non-rotational–material–shape;
d) Shape–rotational or non-rotational–material–size.

Question 15. In Opitz system, the second digit indicates:
a) Type and shape;
b) External shape and external shape elements;
c) External plane surface finishing;
d) Auxiliary hole and gear teeth.

Question 16. Which of the following techniques of grouping does not consider the design and shape aspect?
a) A simple and visual method of cell design;
b) Family formation by classification and codification;
c) Cell formation using production flow analysis;
d) All of the above.

Question 17. Which of the following is basically a material flow simplification technique.
a) A simple and visual method of cell design;
b) Family formation by classification and codification;
c) Cell formation using production flow analysis;
d) All of the above.

Question 18. In cell, manufacturing, the cell size (people) should be:
a) 6 to 12;
b) 10 to 20;
c) 15 to 25;
d) 20 to 30.

Question 19. The following cell formation technique is based on which component shape and design?
  a) Production flow analysis;
  b) Component flow analysis;
  c) Composite component;
  d) Simulation.
Question 20. Which of the following is (are) the benefit(s) of cellular manufacturing.
  a) Job satisfaction;
  b) Job enlargement;
  c) Both (a) and (b);
  d) Job enrichment.

ANSWERS TO MCQs

1 d; 2 b; 3 d; 4 d; 5 b; 6 a; 7 b; 8 d; 9 a; 10 d; 11c; 12 a; 13 d; 14 a; 15 b; 16 c; 17 c; 18 a; 19 c; 20 c.

BIBLIOGRAPHY

Cagliano, R., & Spina, G. (2000). Advanced manufacturing technologies and flexible production, *Journal of Production Management*, 18, 67–106.

Ferguson, S., Siddiqi, A., Lewis, K., & Weck, O. (2007). Flexible and manufacturing systems: Nomenclature and review, *Proceedings of the ASME International Design Engineering Technical Conference & Computers and Information in Engineering Conference*, September, 4–7.

Mehrabi, M.G., Ulsoy, A.G., Koren, Y. (2002). Trends and perspectives in flexible and reconfigurable manufacturing systems, *Journal of intelligent manufacturing*, 13, 135–146.

Saad, S.M. (2003). The reconfiguration manufacturing systems, *Journal of Material Processing Technology*, 34(1), 3–20.

Tilbury, D.M., & Kota, S. (1999). Integrated machine and control design for reconfigurable machine tools, *Journal of Advanced Intelligent Mechatronics*, 1(1), 629–634.

Walczyk, D., Lakshmikanthan, J., & Kirk, D. (1998). Development of reconfigurable tools for forming aircraft body panels, *Journal of Manufacturing Systems*, 1(17), 287–296.

Williams F.P., & Dubosis, J. (2000). Taxonomy of manufacturing flexibility dimension, *Journal of Operational Management*, 18(5), 57–61.

Yigit, A.S., & Ulsoy, A.G. (2002). Dynamic stiffness evaluation for reconfigurable machine tools including weakly non-linear joint characteristics, *Journal of Manufacturing Engineering*, 1(6), 87–101.

4 Rapid Prototyping and Its Techniques

4.1 INTRODUCTION

Rapid prototyping (RP) is a new manufacturing technique that allows for fast fabrication of computer models designed with three-dimensional (3D) computer-aided design (CAD) software. RP is used in a wide variety of industries, from shoe to car manufacturers. This technique allows for fast realization of ideas into functioning prototypes, shortening the design time and leading to successful final products. RP techniques comprise of two general types: additive and subtractive, each of which has its own pros and cons. Subtractive type RP, or a traditional tooling manufacturing process, is a technique in which material is removed from a solid piece of material until the desired design remains. Examples of this type of RP include traditional milling, turning/lathing or drilling to more advanced versions–such as computer numerical control (CNC) and electric discharge machining (EDM). Additive type RP is the opposite of subtractive type RP. Instead of removing material, material is added layer upon layer to build up the desired design such as stereo lithography, fused deposition modeling (FDM) and 3D printing. This chapter will introduce additive type RP techniques: selective laser sintering (SLS), stereo lithography apparatus (SLA), FDM and inkjet-based printing. It will also cover how to properly prepare 3D CAD models for fabrication with RP techniques.

4.1.1 Advantages and Disadvantages of Rapid Prototyping

Subtractive type RP is typically limited to simple geometries, due to the tooling process where material is removed. This type of RP also usually takes a longer time, but the main advantage is that the end product is fabricated in the desired material. Additive type RP, on the other hand, can fabricate most complex geometries in a shorter time and at lower cost. However, additive type RP typically includes extra post-fabrication processes of cleaning, post-curing or finishing. Some of the general advantages and disadvantages of rapid prototyping are:

- Fast and inexpensive method of prototyping design ideas;
- Multiple design iterations;

DOI: 10.1201/9781003476375-4

- Physical validation of design;
- Reduced product development time.

Disadvantages:
- Resolution not as fine as traditional machining (millimeter to sub-millimeter resolution);
- Surface flatness is rough (dependent on material and type of RP).

4.1.2 How Does Rapid Prototyping Work?

Rapid prototyping (RP) includes a variety of manufacturing technologies, although most utilize layered additive manufacturing. However, other technologies used for RP include high-speed machining, casting, molding and extruding. While additive manufacturing is the most common rapid prototyping process, other more conventional processes can also be used to create prototypes.

These processes include:

- Subtractive: whereby a block of material is carved to produce the desired shape using milling, grinding or turning.
- Compressive: whereby a semi-solid or liquid material is forced into the desired shape before being solidified, such as with casting, compressive sintering.

4.2 WHAT ARE THE DIFFERENT TYPES OF RAPID PROTOTYPING?

4.2.1 Photo Polymerization

This fast and affordable technique was the first successful method of commercial 3D printing. It uses a bath of photosensitive liquid which is solidified layer-by-layer using a computer-controlled ultraviolet (UV) light.

4.2.2 Selective Laser Sintering (SLS)

Used for both metal and plastic prototyping, SLS uses a powder bed to build a prototype one layer at a time using a laser to heat and sinter the powdered material. However, the strength of the parts is not as good as with SLA, while the surface of the finished product is usually rough and may require secondary work to finish it.

4.2.3 Fused Deposition Modeling (FDM) or Material Jetting

This inexpensive, easy-to-use process can be found in most non-industrial desktop 3D printers. It uses a spool of thermoplastic filament which is melted inside a printing nozzle barrel before the resulting liquid plastic is laid down layer-by-layer according to a computer deposition program. While the early results generally had poor resolution and were weak, this process is improving rapidly and is fast and cheap, making it ideal for product development.

4.2.4   SELECTIVE LASER MELTING (SLM) OR POWDER BED FUSION

Often known as powder bed fusion, this process is favored for making high-strength, complex parts. Selective laser melting is frequently used in aerospace, automotive, defense and medical industries. This powder bed-based fusion process uses a fine metal powder which is melted in a layer-by-layer manner to build either prototype or production parts using a high-powered laser. Common SLM materials used in RP include titanium, aluminum, stainless steel and cobalt chrome alloys.

4.2.5   LAMINATED OBJECT MANUFACTURING (LOM) OR SHEET LAMINATION

This inexpensive process is less sophisticated than SLM or SLS, but it does not require specially controlled conditions. LOM builds up a series of thin laminates that have been accurately cut with a laser beam or another cutting device to create the CAD pattern design. Each layer is delivered and bonded on top of the previous one until the part is complete.

4.2.6   DIGITAL LIGHT PROCESSING

Similar to SLA, this technique also uses the polymerization of resins which are cured using a more conventional light source than with SLA. While faster and cheaper than SLA, DLP often requires the use of support structures and post-build curing. An alternative version of this is continuous liquid interface production, whereby the part is continuously pulled from a vat, without the use of layers. As the part is pulled from the vat it crosses a light barrier that alters its configuration to create the desired cross-sectional pattern on the plastic.

4.2.7   BINDER JETTING

This technique allows for one or many parts to be printed at one time, although the parts produced are not as strong as those created using SLS. Binder jetting uses a powder bed onto which nozzles spray micro-fine droplets of a liquid to bond the powder particles together to form a layer of the part.

Each layer may then be compacted by a roller before the next layer of powder is laid down and the process begins again. When complete, the part may be cured in an oven to burn off the binding agent and fuse the powder into a coherent part.

4.3   APPLICATIONS

Product designers use this process for rapid manufacturing of representative prototype parts. This can aid visualization, design and development of the manufacturing process ahead of mass production. Originally, rapid prototyping was used to create parts and scale models for the automotive industry although it has since been taken up by a wide range of applications.

THEORETICAL QUESTIONS

Question 1. What is rapid prototyping? Explain the techniques of rapid prototyping.

Question 2. Explain the principle of rapid prototyping with suitable examples.

Question 3. Explain the technologies of rapid prototyping with suitable examples.

Question 4. Explain the applications, merits and demerits of rapid prototyping.

Question 5. Explain the factors influencing accuracy of rapid prototyping parts.

Question 6. Describe the issues related to rapid prototyping with suitable examples.

Question 7. Describe the importance of rapid prototyping related to manufacturing industries.

MULTIPLE CHOICE QUESTIONS

Question 1. Rapid prototyping is a unique technique to ___ for the development of new products.
   a) Reduce the lead time;
   b) To reduce the costs;
   c) Both (a) and (b);
   d) None of the above.

Question 2. The rapid prototyping technique is also known as:
   a) Layered manufacturing;
   b) Free-form fabrication;
   c) Both (a) and (b);
   d) None of the above.

Question 3. In rapid prototyping by laser stereolithography, the resin is cured by:
   a) Laser;
   b) Ultraviolet lamp;
   c) Both (a) and (b);
   d) None of the above.

Question 4. Laser stereolithography requires ___ CAD data in order to create solid models.
   a) One-dimensional;
   b) Two-dimensional;
   c) Three-dimensional;
   d) All of the above.

Question 5. The following is (are) limitations of laser stereolithography:
   a) Metal products cannot be manufactured directly;
   b) Only photocurable resins can be used;
   c) Material strength of resin is worse than common polymer;
   d) All of the above.

Question 6. The following is (are) rapid prototyping processes.
   a) Photopolymer;
   b) Sheet lamination;
   c) Powder sintering;
   d) All of the above.

Question 7. The following is (are) application(s) of rapid prototyping.
    a)   Making wax models for investment casting;
    b)   Making master models for die and model making;
    c)   Mold making for prototype manufacturing;
    d)   All of the above.
Question 8. Dies and molds are unsuitable for:
    a)   Making prototypes;
    b)   Small lot production;
    c)   Both (a) and (b);
    d)   None of the above.
Question 9. Flexible prototype production has been carried out in sheet metal forming with the use of the:
    a)   Turret punch press;
    b)   Laser beam cutting machine;
    c)   Numeric control (NC) press brake;
    d)   All of the above.
Question 10. Which one is NOT related to rapid prototyping definition?
    a)   Layer-by-layer;
    b)   Physical model;
    c)   From 3D CAD data;
    d)   Production line.
Question 11. Which one of these processes is NOT using a laser?
    a)   LOM;
    b)   SLA;
    c)   SLS;
    d)   FDM.
Question 12. Which of the following is a process in the RP cycle?
    a)   Post-processing;
    b)   Transfer to machine;
    c)   Pre-processing;
    d)   All of the above.
Question 13. Which of the processes is available in colors?
    a)   SLA;
    b)   FDM;
    c)   MJM;
    d)   3D printer.
Question 14. What is the full name of SLS?
    a)   Selective laser simulator;
    b)   Sintering laser simulator;
    c)   Selective laser sintering;
    d)   Stereolithography laser sintering.
Question 15. What is the other name of multi jet modeling?
    a)   FDM;
    b)   Poly jet;
    c)   3D printer;
    d)   Extrusion.

Question 16. Which of the following is one of the design process steps?
  a) Build;
  b) Concept;
  c) Pre-processing;
  d) Transfer to machine.

Question 17. Which one of the processes is subtractive prototyping?
  a) Five-axis CNC milling;
  b) Fused deposition modeling;
  c) Multi-jet modeling;
  d) Stereolithography apparatus.

Question 18. In which of these processes is the input material in liquid form?
  a) LOM;
  b) SLS;
  c) FDM;
  d) MJM.

Question 19. Which of the following is NOT the color binder of a 3D printer?
  a) Cyan;
  b) Black;
  c) Magenta;
  d) Yellow.

Question 20. Which of the following is the process of the pre-processing stage?
  a) Remove support;
  b) Checking 3D CAD data;
  c) De-powdering loose material;
  d) Dip in a binder to strengthen the part.

Question 21. Rapid prototyping and additive manufacturing are two terms that refer to fabrication technologies that add layers of material to an existing part or substrate.
  a) True;
  b) False.

Question 22. Machining is never used for rapid prototyping because it takes too long.
  a) True;
  b) False.

Question 23. In the context of rapid prototyping and additive manufacturing, tessellation refers to the process of slicing the CAD model of the part into layers.
  a) True;
  b) False.

Question 24. Ballistic-particle manufacturing is another name for which one of the following RP technologies?
  a) Droplet deposition manufacturing;
  b) Fused-deposition modeling;
  c) Laminated-object manufacturing;
  d) Selective laser sintering.

Question 25. Rapid prototyping technologies are never used to make production parts.
    a)   True;
    b)   False.

ANSWERS TO MCQs

1. c; 2. c; 3. c; 4. c; 5. d;, 6. d; 7. d; 8. c; 9. d; 10. d; 11. d; 12. d; 13. d; 14. c; 15. b; 16. b; 17. a; 18. d; 19. b; 20. b; 21. a; 22. b; 23. b; 24. a; 25. a.

BIBLIOGRAPHY

Arinez, F.J., & Duda, J.W. (2007). Decomposition approach for manufacturing system design, *Journal of Manufacturing System*, 20(6), 371–389.

Bennet, T., Arbel, A., & Seidmann, A. (1993). Towards a new model of sustainable production performance evaluation of flexible manufacturing systems, *IEEE Transactions on Systems*, 14(4), 606–617.

Fricke, M., & Kahyaoglu, Y. (2005). Flexibility and robustness in manufacturing systems, *International Journal of Manufacturing Technology*, 2, 546–558.

Hardt, G., & Gear, T. (1997). Flexibility and its measurement, *Annals of CIRP*, 11(3), 228–230.

Pahl, E., & Burbidge, J.L. (1996). Model development for future manufacturing system, *International Journal of Production Research*, 30(5), 209–1219.

Porter, M. (1996). Towards new manufacturing tools, *International Journal of Production Research*, 31(10), 2403–2414.

Rajan, F., Dubois, D., Rathmill, K., Sethi, S.P., & Stecke, K.E. (2005). Classification of flexible manufacturing systems, *FMS Magazine*, 2, 114–117.

Wiendahl, T., & Nazemetz, J.W. (2007). Performance domain of functional layouts, *Journal of Industrial Engineering*, 34(1), 91–101.

5 Program Logic Controller and Its Communications

5.1 INTRODUCTION

The program logic controller is an industrial digital computer which has been adapted for the control of manufacturing processes, such as assembly lines or any activity that requires high reliability, ease of programming and process fault diagnosis. PLCs can range from small modular devices with tens of inputs and outputs, in a housing integral with the processor, to large rack-mounted modular devices with a count of thousands of I/O, and which are often networked to other PLC and SCADA systems.

They can be designed for many arrangements of digital and analog I/O, extended temperature ranges, immunity to electrical noise and resistance to vibration and impact.

PLCs were first developed in the automobile manufacturing industry to provide flexible, rugged and easily programmable controllers to replace hard-wired relay logic systems. Since then, they have been widely adopted as high-reliability automation controllers suitable for harsh environments. A PLC is an example of a "hard" real-time system, since output results must be produced in response to input conditions within a limited time, otherwise unintended operations will result.

5.2 BASIC FUNCTIONS OF PLC

The most basic function of a programmable controller is to emulate the functions of electro-mechanical relays. Discrete inputs are given a unique address, and a PLC instruction can test if the input state is on or off. Just as a series of relay contacts perform a logical AND function, not allowing current to pass unless all the contacts are closed, so a series of "examine if on" instructions will energize its output storage bit if all the input bits are on. Similarly, a parallel set of instructions will perform a logical OR. In an electro-mechanical relay wiring diagram, a group of contacts controlling one coil is called a "rung" of a "ladder diagram", and this concept is also used to describe PLC logic. Some models of PLC limit the number of series and parallel instructions in one "rung" of logic. The output of each rung sets or clears a storage bit, which may be associated with a physical output address, or which may be an "internal coil" with no physical connection. Such internal coils can be used, for example, as a common element in multiple separate rungs. Unlike physical relays, there is usually

 DOI: 10.1201/9781003476375-5

no limit to the number of times an input, output or internal coil can be referenced in a PLC program.

Some PLCs enforce a strict left-to-right, top-to-bottom execution order for evaluating the rung logic. This is different from electro-mechanical relay contacts, which in a sufficiently complex circuit may either pass current left-to-right or right-to-left, depending on the configuration of surrounding contacts. The elimination of these "sneak paths" is either a bug or a feature, depending on programming style. More advanced instructions of the PLC may be implemented as functional blocks, which carry out some operation when enabled by a logical input and which produce outputs to signal, for example, completion or errors, while manipulating variables internally that may not correspond to discrete logic.

5.3   STEPS OF THE PLC CYCLE

A PLC works in a program scan cycle, where it executes its program repeatedly. The simplest scan cycle consists of three steps:

- Read inputs;
- Execute the program;
- Write outputs.

The program follows the sequence of instructions. It typically takes a timespan of tens of milliseconds for the processor to evaluate all the instructions and update the status of all outputs. If the system contains remote I/O—for example, an external rack with I/O modules—then that introduces additional uncertainty in the response time of the PLC system.

As PLCs became more advanced, methods were developed to change the sequence of ladder execution and subroutines were implemented. This enhanced programming could be used to save scan time for high-speed processes; for example, parts of the program used only for setting up the machine could be segregated from those parts required to operate at higher speed. Newer PLCs now have the option to run the logic program synchronously with the scanning. This means that IO is updated in the background and the logic reads and writes values as required during the logic scanning. Special-purpose modules may be used where the scan time of the PLC is too long to allow predictable performance. Precision timing modules, or counter modules for use with shaft encoders, are used where the scan time would be too long to reliably count pulses or detect the sense of rotation of an encoder. This allows even a relatively slow PLC to still interpret the counted values to control a machine, as the accumulation of pulses is executed by a dedicated module that is unaffected by the speed of program execution.

5.4   HOW DOES A PLC WORK?

The working of a PLC can be easily understood as a cyclic scanning method known as the scan cycle that is shown in Figure 5.1 that is a block diagram of PLC working.

The PLC scan process includes the following steps:

- The operating system starts cycling and monitoring of time.
- The CPU starts reading the data from the input module and checks the status of all the inputs.
- The CPU starts executing the user or application program written in relay-ladder logic or any other PLC-programming language.
- The CPU performs all the internal diagnosis and communication tasks.
- According to the program results, it writes the data into the output module so that all outputs are updated.

This process continues as long as the PLC is in run mode.

5.4.1   PLC Applications

PLCs have a variety of applications and uses, including:

- Process automation plants;
- Glass industry;
- Paper industry;
- Cement manufacturing;
- Boilers in thermal power plants.

5.5   TYPES OF PLCS

The two main types of PLC are fixed/compact PLC and modular PLC.

5.5.1   Compact PLC

Within a single case, there would be many modules, with a fixed number of I/O modules and external I/O cards. Therefore, it does not have the capability to expand the modules. Every input and output would be decided by the manufacturer.

5.5.2   Modular PLC

This type of PLC permits multiple expansions through "modules", and hence is referred to as modular PLC. I/O components can be increased. It is easier to use because each component is independent of the others.

5.5.3   Modular Type PLC

PLCs are divided into three types based on output, namely relay output, transistor output and output PLC. The relay output type is best suited for both AC and DC output devices. Transistor output type PLC uses switching operations and is used inside microprocessors.

According to the physical size, a PLC is divided into mini, micro and nano PLC.

5.6   PLC PROGRAMMING

When using a PLC, it's important to design and implement concepts depending on your particular use case. To do this we first need to know more about the specifics of PLC programming. A PLC program consists of a set of instructions either in textual or graphical form, which represents the logic that governs the process the PLC is controlling. There are two main classifications of PLC programming languages, which are further divided into many sub-classified types.

5.6.1   TEXTUAL LANGUAGE

- Instruction list;
- Structured text.

5.6.2   GRAPHICAL FORM

- Ladder diagrams;
- Function block diagram;
- Sequential function chart.

Due to the simple and convenient features, graphical representation is much preferred to textual languages.

5.6.2.1   Ladder Logic

This is the simplest form of PLC programming. It is also known as "relay logic". The relay contacts used in relay-controlled systems are represented using ladder logic. Figure 5.1 shows an example of a ladder diagram as mentioned below. In the above-mentioned example, two pushbuttons are used to control the same lamp load. When any one of the switches is closed, the lamp will glow. The two horizontal lines are called rungs, and two vertical lines are called rails. Every rung forms an electrical connectivity between Positive rail (P) and Negative rail (N). This allows the current to flow between input and output devices.

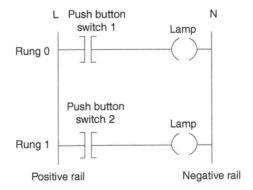

FIGURE 5.1   The ladder diagram.

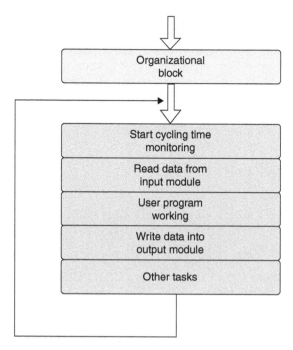

FIGURE 5.2   Block diagram of PLC working.

5.6.2.2   Functional Block Diagrams

A functional block diagram (FBD) is a simple and graphical method to program multiple functions in PLC. PLC Open has been described using FBD in the standard IEC
622-3. A function block is a program instruction unit that, when executed, yields one
or more output values. It is represented by a block as shown below. It is represented
as a rectangular block with inputs entering on left and output lines leaving at the right.
It gives a relation between the state of input and output shown in the functional diagram that is Figure 5.2. The advantage of using FBD is that any number of inputs and
outputs can be used on the functional block. When using multiple input and output,
you can connect the output of one function block to the input of another.

THEORETICAL QUESTIONS

Question 1. Describe the overview of a program logic controller with a suitable
  example.
Question 2. Explain the components of a program logic controller with suitable
  applications.
Question 3. Explain the types of programming device with suitable applications.
Question 4. Explain the operations of a program logic controller with suitable
  diagrams and applications.
Question 5. Explain PLC communications with suitable applications.

Question 6. Explain construction and working of program logic controller with a neat diagram.

Question 7. Discuss the applications, merits and demerits of program logic controller.

MULTIPLE CHOICE QUESTIONS

Question 1. Programmable logic controllers are used in:
a) Manufacturing;
b) Automation;
c) Both (a) and (b);
d) None of the above.

Question 2. What are the components that make the programmable logic controller work?
a) Input and output module;
b) CPU;
c) Power supply;
d) All of the above.

Question 3. The programmable logic controller is classified into how many types:
a) One;
b) Two;
c) Three;
d) Four.

Question 4. In a fixed programmable logic controller:
a) Input is fixed;
b) Output is fixed;
c) Both (a) and (b) are fixed;
d) None of the above.

Question 5. In a modular programmable logic controller:
a) Input is fixed;
b) Output is fixed;
c) Both (a) and (b) are fixed;
d) None of the above.

Question 6. What are the types of programmable logic controllers?
a) Fixed, uniform PLC;
b) Modular, uniform PLC;
c) Fixed and modular PLC;
d) None of the above.

Question 7. The components that make PLC works can be divided into _____ core areas:
a) One;
b) Two;
c) Three;
d) Four.

Question 8. PLC operation _____ checks the status at the input side:
a) Input scan;
b) Output scan;
c) Program scan;
d) None of the above.

Question 9. In PLC operation _____ retrieves the data into an output module:
a) Input scan;
b) Output scan;
c) Program scan;
d) None of the above.

Question 10. Before PLC's were created many industries used:
a) Relays;
b) Capacitors;
c) Resistors;
d) None of the above.

Question 11. Which is the first PLC model?
a) PLC 084;
b) PLC 085;
c) PLC 086;
d) None of the above.

Question 12. CCTV cameras are an example of _____ automation:
a) Building automation;
b) Office automation;
c) Scientific automation;
d) Industrial automation.

Question 13. The control logic in a programmable logic controller can be programmed by:
a) FBD, ladder logic;
b) Sequential logic;
d) Structured text;
e) All of the above.

Question 14. Who invented the programmable logic controller (PLC)?
a) Jonas Wenstrom;
c) Dick Morley;
d) Thomas Davenport;
e) None of the above.

Question 15. Programmable logic controllers are used in:
a) Glass and paper industry;
b) Process automation plants;
c) Cement manufacturing;
d) All of the above.

Question 16. In modular type PLC, the PLCs are classified into:
a) Relay output PLC;
b) Transistor output PLC;
c) Triac output PLC;
d) All of the above.

Question 17. The programmable logic controllers are classified into _____ according to physical size in modular type PLC:
  a) Mini PLC, micro PLC;
  b) Micro PLC, nano PLC;
  c) Nano PLC, mini PLC, macro PLC;
  d) None of the above.

Question 18. The advantages of a PLC are:
  a) Easy maintenance;
  b) Reliability is high;
  c) Small in size;
  d) All of the above.

Question 19. The CPU has:
  a) Memory system;
  b) Processor;
  c) Power supply;
  d) All of the above.

Question 20. The visual programming language is also called:
  a) Relay logic;
  b) Ladder logic;
  c) Controller logic;
  d) All of the above.

Question 21. _____ are the components that are required to change or create a program:
  a) PLC, programming device;
  b) Programming software;
  c) Connector cable;
  d) All of the above.

Question 22. _____ is an example of light automation:
  a) Rocket launching;
  b) Street solar lighting;
  c) Automated bottle filling stations;
  d) Smoke detectors.

Question 23. In the water level storage tank, the manual mode program controls the water level by monitoring the _____ switch input:
  a) Low sensor switch;
  b) High sensor switch;
  c) Both (a) and (b);
  d) None of the above.

Question 24. The sequences are classified into:
  a) One;
  b) Two;
  c) Three;
  d) Four.

Question 25. The PLC internally operates, stores and calculates the value in:
a)   Binary format;
c)   Decimal format;
d)   Octal format;
e)   None of the above.

ANSWERS TO MCQs

1 c; 2 d; 3 b; 4 c; 5 c; 6 c; 7 c; 8 a; 9 b; 10 a; 11a; 12 b; 13 d; 14 b; 15 d; 16 d; 17 c; 18 c; 19 b; 20 d; 21 b; 22 b; 23 a; 24 c; 25 a.

BIBLIOGRAPHY

Farrington, P. A., and & Nazemetz, J. W. (1998). Performance domain of production layouts, *Journal of Industrial Engineering*, 34(1), 91–101.
Karen, B., Landers, R., & Ulsoy, A. (1999). Supervisory machining control design approach and experiments, *Journal of Manufacturing Technology*, 47(1), 301–306.
Maraghy, H.A., Kuzgunkaya, O., & Urbanic, J. (2007). Manufacturing system configurations complexity, *Annals of CIRP*, 5, 445–450.
Rossitza, W., & Lacik, N. (2004). The flexibility of manufacturing system, *International Journal of Production Engineering*, 7(4), 35–457.
Shewchulk, J. P., & Moodie, C.L. (1998). Classification of manufacturing flexibility, *International Journal of Flexibility Manufacturing System*, 10(2), 32–34.
Slack, N. (1987). The flexibility of manufacturing system, *International Journal of Production Management*, 7(3), 35–45.
Spicer, R., Heisel, U., Jovane, F., Moriwaki, T., Pritschow, G., Ulsoy, G., & Van Brussel, H. (2000). Reconfigurable manufacturing systems, *Annals of CIRP*, 3, 527–540.
Zhong, T., Shewchuk, J.P., & Moodie, C.L. (2000). Definition and classification of manufacturing flexibility types and measures, *International Journal of Flexible Manufacturing System*, 10, 325–349.

6 Clamping and Locators, Milling and Grinding Fixtures of Advanced Manufacturing Systems

6.1 INTRODUCTION

A clamping device holds the work piece securely in a jig or fixture against the forces applied over it during on operation. The clamping device should be incorporated into the fixture, proper clamping in a fixture directly influences the accuracy and quality of the work done and production cycle time. The basic requirements of a good clamping device are:

(a) It should rigidly hold the work piece.
(b) The work piece being clamped should not be damaged due to application of clamping pressure by the clamping unit.
(c) The clamping pressure should be enough to overcome the operating pressure applied on the work piece as both pressures act on the work piece in opposite directions.
(d) The clamping device should be unaffected by the vibrations generated during an operation.
(e) It should also be user friendly, e.g., its clamping and releasing should be easy and less time consuming. Its maintenance should also be easy.

6.2 LOCATORS

It is very important to understand the meaning of location before understanding about the jigs and fixtures. The location refers to the establishment of a desired relationship between the workpiece and the jigs or fixture correctness of location which directly influences the accuracy of the finished product. The jigs and fixtures are designed so that all undesirable movements of the work piece can be restricted. Determination of the locating points and clamping of the work piece serve to restrict movements of the component in any direction, while setting it in a particular pre-decided position relative to the jig. Before deciding the locating points, it is advisable to find out all possible degrees of freedom of the work piece. Then some of the degrees of freedom or all of them are restrained by making suitable arrangements. These arrangements are called locators.

DOI: 10.1201/9781003476375-6

6.3 PRINCIPLES OF LOCATIONS

The principle of location is discussed here with the help of a most popular example which is available in any of the books covering jigs and fixtures. It is important that one should understand the problem first. Any rectangular body may have three axes: x-axis; y-axis and z-axis. It can move along any of these axes or any of its movement can be released to these three axes. At the same time the body can also rotate about these axes. The total degree of freedom of the body along which it can move is six. For processing the body, it is required to restrain all the degrees of freedom by arranging suitable locating points and then clamping it in a fixed and required position.

6.4 TYPES OF LOCATORS

Different methods are used for location of a work. The locating arrangement should be decided after studying the type of work, type of operation and degree of accuracy required. The volume of mass production to be done also matters. Different locating methods are described below.

6.4.1 FLAT LOCATOR

Flat locators are used for location of flat machined surfaces of the component. Three different examples which can serve as a general principle of location are described here for flat locators. In this case an undercut is provided at the bottom where two perpendicular surfaces intersect each other. This is made for swarf clearance. There is no need to make undercut for swarf clearance. The button can be adjusted to decide very fine location of the work piece.

6.4.2 CYLINDRICAL LOCATORS

This is used for locating components having drilled holes. The cylindrical component to be located is gripped by a cylindrical locator fitted to the jig's body and inserted in the drilled hole of the component. The face of the jig's body around the locator is undercut to provide space for swarf clearance.

6.4.3 CONICAL LOCATOR

This is used for locating the work pieces having a cylindrical hole in the work piece. The work piece is located by supporting it over the conical locator inserted into the drilled hole of the work piece. A conical locator is considered as superior, as it has a capacity to accommodate a slight variation in the hole diameter of the component without affecting the accuracy of location. Degree of freedom along z-axis can also be restrained by putting a template over the work piece with the help of screws.

6.4.4 JACK PIN LOCATOR

A jack pin locator is used for supporting rough work pieces from the button. The height of the jack pin is adjustable to accommodate the work pieces having variation in their surface texture. This is a suitable method to accommodate the components which are rough and un-machined.

6.4.5 VEE LOCATORS

This is a quick and effective method of locating the work piece with the desired level of accuracy. This is used for locating circular and semi-circular type of work pieces. The main part of the locating device is a vee-shaped block which is normally fixed to the jig. This locator can be of two types: fixed vee locator and adjustable vee locator. The fixed type locator is normally fixed on the jig and the adjustable locator can be moved axially to provide a proper grip of the vee band to the work piece.

6.5 CLAMPING PRINCIPLES

- The clamp should firmly hold the work piece without distorting it.
- The clamp should overcome the maximum possible force exerted on the work piece by using minimum clamping force.
- The clamp should be easy to operate.
- Vibrations should tighten the cams and wedges in the clamp design (if any) and not loosen them.

6.6 TYPES OF CLAMPING

The different type of clamping are:

- Mechanical actuation clamps;
- Pneumatic and hydraulic clamps;
- Vacuum clamping;
- Magnetic clamping;
- Electrostatic clamping;
- Non-mechanical clamping;
- Special clamping operations.

While designing for clamping the following factors essentially need to be considered:

- Clamping should be easy, quick and consistently adequate.
- Clamping should be such that it is not affected by vibration or heavy pressure.
- The way of clamping and unclamping should not hinder loading and unloading the blank in the jig or fixture.
- Clamp and clamping force must not damage the work piece.
- Clamping operations should be very simple and quick acting when the jig or fixture is to be used more frequently and for large volumes of work which move by slide or slip or tend to do so during applying clamping forces, should be avoided.
- The clamping system should comprise fewer parts for ease of design, operation and maintenance.

- The wearing parts should be hard and easily replaceable.
- The clamping force should act on heavy parts and against supporting and locating surfaces.
- The clamping force should be away from the machining thrust forces.
- The clamping method should be fool proof and safe.
- Clamping must be reliable but also inexpensive.

6.7   GENERAL METHODS OF LOCATING

6.7.1   LOCATING BLANKS FOR MACHINING IN LATHES

In lathes, where the job rotates, the blanks are located by fitting into a self-centring chuck–fitting to four independent jaw chuck and dead center in self–centring collets in between live and dead centers by using a mandrel fitted into the head stock–spindle fitting in a separate fixture which is properly clamped on a driving plate which is coaxially fitted into the lathe spindle.

6.7.2   LOCATING FOR MACHINING IN OTHER THAN LATHES

In machine tools like drilling machines, boring machines, milling machines, planing machines, broaching machines and surface grinding machines, the job remains fixed on the bed or worktable of the machine tools. Fixtures are mostly used in the afore-said machine tools and jigs specially for drilling, reaming, etc., for batch production.

6.7.3   LOCATING BY FLAT SURFACES

Locating jobs by their flat surfaces using various types of flat ended pins and buttons.

6.7.4   LOCATING BY HOLES

In several cases, work pieces are located by premachined (drilled, bored or pierced) holes, such as locating by two holes.

6.7.5   LOCATING ON MANDREL OR PLUG

Ring or disc type work pieces are conveniently located on mandrels or single plugs. However, there may be several other ways of locating depending upon the machining conditions and requirements.

6.8   BASIC PRINCIPLES OR RULES TO BE FOLLOWED WHILE DESIGNING OR PLANNING FOR SUPPORTING

- Supporting should be provided at least at three points.
- Supporting elements and systems have to be strong enough and rigid to prevent deformation due to clamping and cutting forces.
- Unsupported spans should not be large enough to cause sagging.

- Supporting should keep the blank in a stable condition under the forces applied.
- Supporting large flat areas a proper recess is to be provided, for better and stable support.
- Round or cylindrical work pieces should be supported (along with locating) on a strong vee block of suitable size.
- Heavy work pieces with pre-machined bottom surface should be supported on wide flat areas, otherwise on flat ended strong pins or plugs.
- If more than three pins are required for supporting large work pieces, then the additional supporting pins are to be spring loaded or adjustable.
- Additional adjustable supporting pins need to be provided.

6.9   INTRODUCTION TO MILLING FIXTURES

A milling fixture is a work holding device which is firmly clamped to the table of the milling machine. It holds the work piece in the correct position as the table movement carries it past the cutter or cutters.

6.9.1   ESSENTIALS OF MILLING FIXTURES

6.9.1.1   Base

A heavy base is the most important element of a milling fixture. It is a plate with a flat and smooth under face. The complete fixture is built up from this plate. Keys are provided on the under face of the plate which are used for easy and accurate aligning of the fixture on the milling machine table, by inserting them into the T-slot in the table. These keys are usually set in keyways on the under face of the plate and are held in place by a socket head cap screw for end key. The fixture is fastened to the machine table with the help of two T-bolts engaging in T-slots of the work table.

After the fixture has been securely clamped to the machine table, the work piece which is correctly located in the fixture has to be set in correct relationship to the cutters. This is achieved by the use of setting blocks and feeler gauges. The setting block is fixed to the fixture. Feeler gauges are placed between the cutter and reference planes on the setting block so that the correct depth of the cut and correct lateral setting is obtained. The block is made of hardened steel with the feeler surfaces grooved. In a correct setting, the cutter should clear the feeler surfaces by at least 0.08 cm to avoid any damage to the block when the machine table is moved back to unload the fixture. The thickness of the feeler gauge to be used should be stamped on the fixture base near the setting block, as shown in Figure 6.1.

6.10   DESIGN PRINCIPLE FOR MILLING FIXTURE

The design principles of location and clamping apply for milling fixtures are:•Pressure of cut should always be against the solid part of the fixture;

- Clamps should always operate from the front of the fixture;
- The work piece should be supported as near the tool thrust as possible.

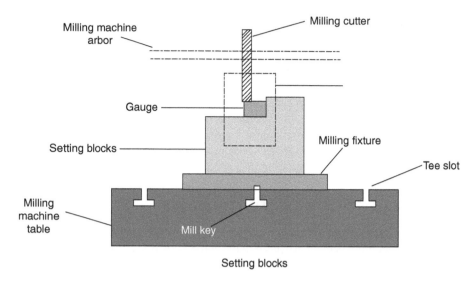

FIGURE 6.1   Detailed features of setting block.

6.11   GRINDING FIXTURES

The work holding devices for grinding operations will depend upon the type of the grinding operation and the machine used.

6.11.1   FIXTURE FOR EXTERNAL GRINDING

A mandrel is the most common fixture used for grinding external surfaces of the work piece, a mandrel is hardened and is held between centers of a machine. The mandrel is used for internal chucking or round work pieces with bores. The work piece is located and held on the mandrel with the help of the bore so that the external surface may be machined truly concentric to the bore.

6.11.2   TAPER MANDREL

In this type of mandrel, the outer chucking surface is given a slender taper of about 0.5 mm per meter.

6.11.3   STRAIGHT MANDREL

It differs from the taper mandrel in that it has a straight tapered chucking surface.

6.12   FIXTURES FOR INTERNAL GRINDING

For grinding internal surfaces of a simple circular work piece, the chuck may be used as a standard work-holding device. If required special jaws can be provided for the chuck. However, for many components special fixtures may have to be made which are designed on the same lines as the lathe fixtures.

6.13   FIXTURES FOR SURFACE GRINDING

The work piece can be held for machining on a surface grinder in the following ways:

1. It may be clamped directly to the machine table or to an angle plate and so on.
2. It may be held in a vice.
3. The work piece may be held by means of a magnetic chuck or a vacuum chuck. Here the work piece is held without any mechanical clamping.
4. The work piece may be held in a special fixture.

6.14   METHODS OF CUTTING OPERATION

6.14.1   ORTHOGONAL CUTTING PROCESS

Orthogonal cutting occurs when the major cutting edge of the tool is presented to the work piece perpendicular to the direction of the feed motion. Orthogonal cutting is shown in Figure 6.2.

6.14.2   OBLIQUE CUTTING PROCESS

Oblique cutting occurs when the major edge of the cutting tool is presented to the work piece at an angle which is not perpendicular to the direction of the feed motion, its diagram shows that chip removals are the continuous type shown in Figure 6.2.

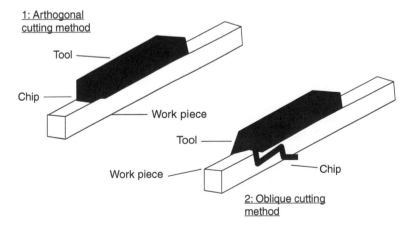

FIGURE 6.2   Shows the chip removal process.

THEORETICAL QUESTIONS

Question 1. Explain different types of clamps with neat and clean diagrams.

Question 2. Explain different types of locators with neat diagrams.

Question 3. Explain the principle of clamps and locators.

Question 4. Explain locating methods with suitable applications.

Question 5. Explain non-conventional clamping methods with a suitable diagram.

Question 6. Explain the overview of milling fixtures with neat and clean diagrams.

Question 7. Explain the types of milling fixtures with neat and clean diagrams and also mention their applications.

Question 8. Describe the essentials of milling fixtures with suitable examples.

Question 9. Explain the types of cutter with neat diagrams and also mention their applications.

Question 10. Describe the concept of direction of feed with suitable examples.

Question 11. Describe the grinding fixtures with a neat diagram and mention its applications.

Question 12. Explain the types of grinding fixtures with neat diagrams and also mention its merits, demerits and applications.

MULTIPLE CHOICE QUESTIONS

Question 1. The work piece which is so arranged in the body of jig and fixture, considering the position of the tool, that it and other similar work pieces can occupy only one position is known as:
    a)   Location;
    b)   Clamping;
    c)   Jigs;
    d)   None of the above.

Question 2. What are the locating and clamping principles?
    a)   Reduce operation time;
    b)   Increase productivity;
    c)   Reduce the cost of production;
    d)   All of the above.

Question 3. Proper location of clamping devices is to be done by adopting _____:
    a)   3-2-1 method;
    b)   4-2-2 method;
    c)   4-2-1 method;
    d)   1-2-1 method.

Question 4. The 3-2-1 principle is also known as:
    a)   Six-point location;
    b)   Principle of location;
    c)   x-x-z location of principle;
    d)   All of the above.

Question 5. Which fixtures are used for machining parts which must have machined details evenly spaced?

a) Profile fixtures;
b) Duplex fixtures;
c) Indexing fixtures;
d) None of the above.

Question 6. V-blocks (Vee locators) are used for clamping as well as locating when faces are inclined up to:

a) 30 degrees;
b) 12 degrees;
c) 9 degrees;
d) 3 degrees.

Question 7. What should be the tolerance on a jig?

a) 10–15% of the tolerance on the job;
b) 20–50% of the tolerance on the job;
c) 05–10% of the tolerance on the job;
d) 50–100% of the tolerance on the job.

Question 8. What fixture is used as a production device?

a) Hold work piece;
b) Locate work piece;
c) Hold and locate the work piece;
d) Rotate work piece.

Question 9. A drill jig bushing is generally made of:

a) Cast iron;
b) Tool steel;
c) Cast steel;
d) Mild steel.

Question 10. Which among the following locators is best suitable for location of a round shaped job:

a) "V" locator;
b) Pin type locator;
c) Adjustable stop locator;
d) Wedge type locator.

Question 11. Which jig is used for drilling on thin sheet:

a) Table jig;
b) Post jig;
c) Sandwich jig;
d) Solid jig.

Question 12. Which of the following is not a holding device in a milling machine:

a) "V" block;
b) Tail stock;
c) Fixture;
d) Collets.

Question 13. What part carries the tool head in a shaping machine:
    a) Table;
    b) Column;
    c) Cross rail;
    d) Ram.
Question 14. Which of the following is used to guide the tool and hold the job in mass production:
    a) Jig;
    b) Fixture;
    c) Housing;
    d) Gauge.
Question 15. The size of the milling machine is given by the dimension and maximum travel in three directions of which part:
    a) Cutter;
    b) Arbor;
    c) Column;
    d) Worktable.
Question 16. Which device is used in the shaping machine to hold the cylindrical work piece:
    a) Plummer block;
    b) Square block;
    c) "U" block;
    d) "V" block.
Question 17. Which of the following jigs is made from a single block of steel to drill small work pieces:
    a) Box jig;
    b) Channel jig;
    c) Angle jig;
    d) Solid jig.
Question 18. Jigs and fixtures are used for:
    a) Mass production;
    b) Identical production;
    c) Both (a) and (b);
    d) None of the above.
Question 19. The use of jigs and fixtures:
    a) Facilitates deployment of less skilled labor for production;
    b) Eliminates pre-machining operations like marking, measuring, laying out, etc.
    c) Reduces manual handling operations;
    d) All of the above.
Question 20. The following is(are) the function(s) of a jig:
    a) Holding;
    b) Locating;
    c) Guiding;
    d) All of the above.

Question 21. A fixture does not:
   a)   Hold the workpiece;
   b)   Locate the workpiece;
   c)   Guide the tool;
   d)   All of the above.

Question 22. Jigs are not used in:
   a)   Drilling;
   b)   Reaming;
   c)   Tapping;
   d)   Milling.

Question 23. Fixtures are used in:
   a)   Milling;
   b)   Shaping;
   c)   Turning;
   d)   All of the above.

Question 24. The principle of _____ states that: "In order to achieve the maximum accuracy in location the locating points should, therefore, be placed as far apart from one another as it is possible".
   a)   Six-point location;
   b)   Least points;
   c)   Extreme positions;
   d)   Mutually perpendicular planes.

Question 25. The following holds the work piece securely in a jig or fixture against the cutting forces:
   a)   Locating device;
   b)   Clamping device;
   c)   Guiding device;
   d)   Indexing device.

Question 26. The following is a quick acting clamp:
   a)   Hinged clamp;
   b)   Cam operated clamp;
   c)   Bridge clamp;
   d)   Edge clamp.

Question 27. The following material is commonly used for making locating and clamping devices:
   a)   High carbon steel;
   b)   Low carbon steel;
   c)   High speed steel;
   d)   Die steel.

Question 28. The following type of jig is used for machining in more than one plane:
   a)   Template jig;
   b)   Plate type jig;
   c)   Open type jig;
   d)   Box type jig.

Question 29. The following type of jig suits best for drilling of holes in hollow cylindrical components, with relatively smaller outside and inside diameters, such as bushes:
  a) Solid type jig;
  b) Pot type jig;
  c) Box type jig;
  d) Open type jig.

Question 30. The following type of jig is used to drill a series of equidistant holes along a circle:
  a) Index jig;
  b) Plate type jig;
  c) Open type jig;
  d) Pot type jig.

Question 31. If a grinding wheel is specified as "A 46 K 5 B 17", the grain size of the wheel will be:
  a) Coarse;
  b) Medium;
  c) Fine;
  d) Very fine.

Question 32. If a grinding wheel is specified as "C 8 K 5 B 17", The grain size of a wheel will be:
  a) Coarse;
  b) Medium;
  c) Fine;
  d) Very fine.

Question 33. If a grinding wheel is specified as "A 600 K 5 B 17", the grain size of a wheel will be:
  a) Coarse;
  b) Medium;
  c) Fine;
  d) Very fine.

Question 34. Which of the following grinding wheels will have a fine grain size:
  a) A 46 K 5 B 17;
  b) C 600 K 5 B 17;
  c) C 8 K 5 B 17;
  d) A 80 K 5 B 17.

Question 35. Which of the following grinding wheels will have a fine grain size?
  a) A 46 K 5 B 17;
  b) C 600 K 5 B 17;
  c) C 8 K 5 B 17;
  d) A 80 K 5 B 17.

Question 36. Which of the following specified grinding wheels will have an aluminum oxide abrasive:
  a) Z 46 K 5 B 17;
  b) C 600 K 5 B 17;

    c)   C 8 K 5 B 17;
    d)   A 80 K 5 B 17.
Question 37. Which of the following specified grinding wheels will have a zirconia abrasive:
    a)   Z 46 K 5 B 17;
    b)   C 600 K 5 B 17;
    c)   C 8 K 5 B 17;
    d)   A 80 K 5 B 17.
Question 38. Which of the following specified grinding wheels will have a silicon carbide abrasive?
    a)   Z 46 K 5 B 17;
    b)   C 600 K 5 B 17;
    c)   A 8 K 5 B 17;
    d)   A 80 K 5 B 17.

ANSWERS TO MCQs

1 a; 2 d; 3 d; 4 a; 5 c; 6 c; 7 b; 8 c; 9 b; 10 a; 11 c; 12 b; 13 d; 14 a; 15 d; 16 d; 17 d; 18 c; 19 d; 20 d; 21 c; 22 d; 23 d; 24 c; 25 b; 26 b; 27 a; 28 d; 29 a; 30 a; 31 b; 32 a; 33 d; 34 d; 35 c; 36 d; 37 a; 38 b.

BIBLIOGRAPHY

Cho, M.J., & Kim, M. (1996). Enabling technologies of agile manufacturing and its related activities in Korea, *Computers Industrial Engineering*, 3, 323–334.

Dowlatshahi, S., & Cao, Q. (2005). The relationships among virtual enterprise, information technology and business performance in agile manufacturing: An industry perspective, *International Journal of Operational Research*, 2, 234–245.

Lander, R., & Minko, B.K. (2001). Reconfigurable machine tools, *Annals of CIRP*, 49(1), 69–74.

Mehrabi, M.G., Ulsoy, A.G., & Koren, Y. (2000). Reconfigurable Manufacturing systems and their enabling technologies, *International Journal of Manufacturing Technology and Management*, 2, 113–130.

Naylor, J.B., Naim, M.M., & Berry, D. (1999). Leagility: Interfacing the lean and agile manufacturing paradigm in the total supply chain, *International Journal of Production Economics*, 62, 107–111.

Nohria, N., & Berkley, J.D. (1994). An action perspective: The cruz of the new management, *California Management Review*, 3, 70–92.

Sethi, A.K., & Kota, S.P. (1990). Flexibility in manufacturing system, a survey, *International Journal of Flexible Manufacturing System*, 2(3), 89–92.

Sharp, J.M., Irani, Z., & Desai, S. (1999). Working towards agile manufacturing in the UK industry, *International Journal of Production Economics*, 6, 155–169.

7 Artificial Intelligence and Flexible Manufacturing Systems

7.1 INTRODUCTION

A flexible manufacturing system (FMS) is a production method that is designed to easily adapt to changes in the type and quantity of the product being manufactured. Machines and computerized systems can be configured to manufacture a variety of parts and handle changing levels of production.

A flexible manufacturing system (FMS) can improve efficiency and thus lower a company's costs. Flexible manufacturing can also be a key component of a make-to-order strategy that allows customers to customize the products they want.

Such flexibility can come with higher upfront costs. Purchasing and installing the specialized equipment that allows for such customization may be costly compared with more traditional systems.

7.2 HOW FLEXIBLE MANUFACTURING SYSTEMS WORK

The concept of flexible manufacturing was developed by scientist Jerome Lemelson, an American industrial engineer and inventor who filed a number of related patents in the early 1950s. His original design was a robot-based system that could weld, rivet, convey and inspect manufactured goods.

A flexible manufacturing system may include a configuration of interconnected processing workstations with computer terminals that process the end-to-end creation of a product, from loading/unloading functions to machining and assembly, to storing, quality testing and data processing. The system can be programmed to run a batch of one set of products in a particular quantity and then automatically switch over to another set of products of another quantity.

7.2.1 DISADVANTAGES OF FMS

1. Higher upfront costs.
2. Greater time required to design the system specifications for a variety of future needs.

DOI: 10.1201/9781003476375-7

3. There is also a cost associated with the need for specialized technicians to run, monitor and maintain the FMS. Advocates of FMS maintain that the increase in automation typically results in a net reduction in labor costs.

7.2.2 ADVANTAGES OF A FLEXIBLE MANUFACTURING SYSTEM

1. Less scrap;
2. Fewer workstations;
3. Quicker changes of tools, dies and stamping machinery;
4. Reduced downtime;
5. Improved quality through better control over it;
6. Reduced labor costs due to increase in labor productivity;
7. Increase in machine efficiency;
8. Reduced work-in-process inventories;
9. Increased capacity;
10. Increased production flexibility;
11. Faster production;
12. Lower cost/unit;
13. Increased system reliability;
14. Adaptability to CAD/CAM operations.

7.3 COMPONENTS OF A FLEXIBLE MANUFACTURING SYSTEM

The components of a flexible manufacturing system are:

1. Pallet and fixtures;
2. Machining centers;
3. Inspection equipment;
4. Chip removal system;
5. In process storage inventories;
6. Material handling system.

7.3.1 PALLET AND FIXTURES

The functional component which allows for integration of machines, material handling and in process storage is to use palletized parts. The palletized part is a steel disk with slots on the surface. These slots are used to fasten the fixture to the pallet.

7.3.2 MACHINING CENTERS

The machining center consists of column, worktable and tool storage. The machining center has flexibility in performing many different types of machining operations such as drilling.

7.3.3 INSPECTION EQUIPMENT

Monitoring the quality of operation in an FMS is necessary. This is usually done through: 1. Coordinate measuring machine; 2. Probing machine center; 3. Robots.

7.3.4 CHIP REMOVAL SYSTEM

Chips are the pieces of metal which have been removed from the work piece. Two methods of removing these from work areas are chip conveyor to the collection box and an in-floor flume system with a centralized collection area.

7.3.5 IN PROCESS STORAGE INVENTORIES

Due to the lack of perfect information, some buffering is needed between the handling system and machine. The buffering of parts is called in process inventories.

7.3.6 MATERIAL HANDLING SYSTEM

The primary purpose of work handling equipment was to transfer pallet and work piece between the loading and unloading stations. It is now also used to handle tooling and to integrate various forms of storage into systems.

7.4 COMPARISONS OF RMS AND FMS MANUFACTURING SYSTEMS

The reconfigurable manufacturing system will not be more expensive than a flexible manufacturing system or even dedicated transfer lines. Reconfigurable manufacturing systems and flexible manufacturing systems (FMS) have different goals. FMS aims at increasing the variety of parts produced. RMS aims at increasing the speed of responsiveness to markets and customers. RMS is also flexible, but only to a limited extent——its flexibility is confined to only that necessary to produce a part family. This is the customization characteristic, which is not the general flexibility that FMS offers. The advantages of customized flexibility are faster throughput and higher production rates. Other important advantages of RMS are rapid scalability to the desired volume and convertibility, which are obtained with reasonable cost to manufacturers. The best application of an FMS is found in the production of small sets of products. The dedicated transfer lines typically have high capacity but limited functionality. They are cost effective as long as they produce a few part types and demand exceeds supply, but with saturated markets and the increasing pressure of global competition, there are situations where dedicated transfer lines do not operate at their full capacity, which creates a loss. Flexible systems, on the other hand are built with all the flexibilities and functionality available in some cases, with some that may not be needed at installation time. In terms of capacity and functionality, the reconfigurable manufacturing system always lies between the dedicated manufacturing system and flexible manufacturing system. The RMS will be designed by the use of reconfigurable hardware and software, such that its capacity can be changed over time, and unlike the other manufacturing systems, it does not have fixed hardware or software. It is clear that current trends in reconfigurable software and reconfigurable hardware are the key enabling technologies of reconfigurable manufacturing systems.

THEORETICAL QUESTIONS

Question 1. Explain the features of FMS with suitable examples.

Question 2. Describe artificial intelligence with suitable examples.

Question 3. Describe the components of FMS with suitable applications.

Question 4. Describe the merits, demerits and applications of FMS.

Question 5. Compare FMS and RMS with suitable examples.

Question 6. Describe the role of artificial intelligence in flexible manufacturing with suitable examples.

MULTIPLE CHOICE QUESTIONS

Question 1. A flexible manufacturing system (FMS) is a group of ___, interconnected by a central control system.

    a) Special purpose machines;

    b) Numerically-controlled machine tools;

    c) General purpose machines;

    d) Any of the above.

Question 2. A flexible manufacturing system (FMS) is generally limited to firms involved in:

    a) Mass production;

    b) Batch production;

    c) Both (a) and (b);

    d) None of the above.

Question 3. The following kind(s) of manufacturing systems fall within the FMS spectrum:

    a) Components assembly;

    b) Forming system;

    c) Both (a) and (b);

    d) None of the above.

Question 4. The following is (are) the component(s) of generic FMS:

    a) Material handling system;

    b) Set of workstations having machine tools;

    c) Central control computer;

    d) All of the above.

Question 5. The machines in FMS typically perform:

    a) Milling;

    b) Drilling;

    c) Turning;

    d) All of the above.

Question 6. The following is (are) advantage(s) of a flexible manufacturing system.
  a) Fewer workstations;
  b) Work-in-process inventory reduced;
  c) Reduced downtime;
  d) All of the above.
Question 7. The following is true for a flexible manufacturing system (FMS).
  a) It can handle a narrow range of part varieties;
  b) It requires a longer planning and development period;
  c) Equipment utilization is not always as high as one would expect;
  d) All of the above.
Question 8. The flexibility generally considered in FMS is:
  a) Routing flexibility;
  b) Machine flexibility;
  c) Both (a) and (b);
  d) None of the above.
Question 9. Flexibility in manufacturing means the ability to deal with:
  a) Variation in parts assembly and variations in the process sequence;
  b) Change the design of a certain product;
  c) Change the production volume;
  d) All of the above.
Question 10. The FMS data traffic consists of ___ and ___:
  a) Large files, short messages;
  b) Large files, long messages;
  c) Small files, short messages;
  d) Small files, long messages.
Question 11. Flexible manufacturing systems (FMS) are reported to have a number of benefits. Which is NOT a reported benefit of FMS:
  a) More flexible than the manufacturing systems they replace;
  b) Lead time and throughput time reduction;
  c) Increased quality;
  d) Increased utilization.
Question 12. Which materials-processing technology gives the advantage of precision, accuracy and optimum use of cutting tools, which maximize their life and provide higher labor productivity:
  a) Industrial robots;
  b) Computer-integrated manufacturing (CIM);
  c) Flexible manufacturing systems (FMS);
  d) NC (and CNC) machine tools.
Question 13. What do flexible manufacturing systems (FMS) do?
  a) Move and manipulate products, parts or tolls;
  b) Move materials between operations;
  c) Coordinate the whole process of manufacturing and manufacture a part, component or product;
  d) Completely manufacture a range of components without significant human intervention during the processing.

Question 14. What is the type in which the range or universe of part styles that can be produced on the system:
a) Mixed flexibility;
b) Production flexibility;
c) Volume flexibility;
d) Product flexibility.

Question 15. FMS can be classified based on:
a) Kinds of operation they perform;
b) Number of machines;
c) Level of flexibility;
d) All of the above.

Question 16. One of the classifications of FMS based on the number of machines in the system is:
a) Flexible manufacturing cell;
b) Random-order FMS;
c) Dedicated FMS;
d) None of the above.

Question 17. VMC and HMC can be categorized as:
a) Workstations;
b) Load and unload stations;
c) Fixtures;
d) Workpiece transport equipment.

Question 18. The full form of AGV is:
a) Automated guard vehicle;
b) Automated guided vehicle;
c) Automated grinding vehicle;
d) Automated ground vehicle.

Question 19. AGVs can be used as:
a) Workstations;
b) Load and unload stations;
c) Workpiece transport equipment;
d) Pallets.

Question 20. _____ are used to locate parts precisely on pallets.
a) Tools;
b) Fixtures;
c) Load and unload stations;
d) Workstations.

Question 21. What does FMS stand for:
a) Finite machining system;
b) Flexible machining system;
c) Finite manufacturing system;
d) Flexible manufacturing system.

Question 22. Which of the following are the types of flexible manufacturing system?
    a) Sequential, random, dedicated, engineered and modular;
    b) Progressive, loop, open field and ladder;
    c) Workstations, material handling and storage, computer control;
    d) Sort, set in order, shine, standardize and sustain.

Question 23. What are the types of FMS layouts:
    a) Crossed, circular, asymmetric, parabolic;
    b) In line, loop, open field and robot centered;
    c) Square, pyramidal, circular and hyperbolic;
    d) Triangular, trapezoidal and angular.

Question 24. What are the tests conducted for checking flexibility:
    a) Raw material, operator and dye penetrant test;
    b) Part variety, schedule change, error recovery and new part test;
    c) Reliability, processing and quality test;
    d) Non-destructive, material packaging and corrosion resistant test.

Question 25. Which of the following is the process of converting raw materials into finished products:
    a) Trading;
    b) Manufacturing;
    c) Testing;
    d) Inspecting.

Question 26. What is the full form of "AI":
    a) Artificially intelligent;
    b) Artificial intelligence;
    c) Artificially intelligence.
    d) Advanced intelligence.

Question 27. What is artificial intelligence:
    a) Artificial intelligence is a field that aims to make humans more intelligent;
    b) Artificial intelligence is a field that aims to improve security;
    c) Artificial intelligence is a field that aims to develop intelligent machines;
    d) Artificial intelligence is a field that aims to mine data.

Question 28. Who is the inventor of artificial intelligence:
    a) Geoffrey Hinton;
    b) Andrew Ng;
    c) John McCarthy;
    d) Jürgen Schmidhuber.

Question 29. Which of the following is a branch of artificial intelligence:
    a) Machine learning;
    b) Cyber forensics;
    c) Full-stack developer;
    d) Network design.

Question 30. What is the goal of artificial intelligence:
    a) To solve artificial problems;
    b) To extract scientific causes;

c) To explain various sorts of intelligence;

d) To solve real-world problems.

Question 31. Which of the following is an application of artificial intelligence:

  a) It helps to exploit vulnerabilities to secure the firm;

  b) Language understanding and problem-solving (text analytics and NLP);

  c) Easy to create a website;

  d) It helps to deploy applications on the cloud.

Question 32. In how many categories is the process of artificial intelligence categorized:

  a) Five categories;

  b) Processes are categorized based on the input provided;

  c) Three categories;

  d) Process is not categorized.

Question 33. Based on which of the following parameters is artificial intelligence categorized:

  a) Based on functionality only;

  b) Based on capabilities only;

  c) Based on capabilities and functionality;

  d) It is not categorized.

Question 34. Which of the following is a component of artificial intelligence:

  a) Learning;

  b) Training;

  c) Designing;

  d) Puzzling.

Question 35. What is the function of an artificial intelligence "agent":

  a) Mapping of goal sequence to an action;

  b) Work without the direct interference of people;

  c) Mapping of precept sequence to an action;

  d) Mapping of environment sequence to an action.

Question 36. Which of the following is not a type of artificial intelligence agent:

  a) Learning AI agent;

  b) Goal-based AI agent;

  c) Simple reflex AI agent;

  d) Unity-based AI agent.

Question 37. What is the name of the artificial intelligence system developed by Daniel Bobrow:

  a) Program known as BACON;

  b) System known as STUDENT;

  c) Program known as SHRDLU;

  d) System known as SIMD.

Question 38. What is the function of the system Student:

  a) Program that can read algebra word problems only;

  b) System which can solve algebra word problems but not read.

  c) System which can read and solve algebra word problems.

  d) None of the above.

Question 39. Which of the following is not an application of artificial intelligence:
a) Face recognition system;
b) Chatbots;
c) LIDAR;
d) DBMS.

Question 40. Which of the following machines requires input from humans but can interpret the outputs themselves:
a) Actuators;
b) Sensor;
c) Agents;
d) AI system.

Question 41. How many informed search methods are there in artificial intelligence:
a) Four;
b) Three;
c) Two;
d) One.

Question 42. The total number of proposition symbols in AI is:
a) Three proposition symbols;
b) One proposition symbol;
c) Two proposition symbols;
d) No proposition symbols.

Question 43. The total number of logical symbols in AI are:
a) There are three logical symbols;
b) There are five logical symbols;
c) Number of logical symbols are based on the input;
d) Logical symbols are not used.

Question 44. Which of the following are the approaches to artificial intelligence:
a) Applied approach;
b) Strong approach;
c) Weak approach;
d) All of the above.

Question 45. A face recognition system is based on which type of approach:
a) Weak AI approach;
b) Applied AI approach;
c) Cognitive AI approach;
d) Strong AI approach.

ANSWERS TO MCQs

1 b; 2 b; 3 c; 4 d; 5 d; 6 d; 7 d; 8 c; 9 d; 10 a; 11 a; 12 d; 13 d; 14 b; 15 d; 16 a; 17 a; 18 b; 19 c; 20 b; 21 d; 22 a; 23 b; 24 b; 25 b; 26 b; 27 c; 28 c; 29 a; 30 c; 31 b; 32 c; 33 c; 34 a; 35 c; 36 d; 37 b; 38 c; 39 d; 40 d; 41 a; 42 c; 43 b; 44 d; 45 d.

BIBLIOGRAPHY

Alsafi, Y., & Vyatkin, V. (2010). Ontology based reconfigurable agent for intelligent mechatronic system in flexible manufacturing, *Journal of Robotics and Computer Integrated Manufacturing*, 26, 381–391.

Baldwin, C.Y., & Clark, K.B. (1997). Managing in the age of modularity, *Harvard Business Review*, 5, 84–93.

ElMaraghy, H.A. (2007). Flexible and reconfigurable manufacturing systems paradigms, *International Journal of Flexible Manufacturing System*, 17(4), 261–276.

Li, J., Dai, X., Meng, Z., Dou, J., & Guan, X. (2009). Rapid design and reconfigurable of Petri net models for reconfigurable manufactuiriung cells with improved net rewriting systems and activities diagrams, *Journal of Computer and Industrial Engineering*, 5, 1431–1451.

Monila, A., Rodriguez, C., Ahuett, H., Cortes, J., Ramirez, G., & Jimenez, S. (2008). Next generation manufacturing system: key research issues in developing and integrating reconfigurable and intelligent machines, *International Journal of Computer Integrated Manufacturing*, 18(7), 525–536.

Raj, T., Shankar, R., & Suhaib, M. (2008). An ISM approach for modeling the enablers of flexible manufacturing system: the case for India, *International Journal of Production Research*, 46(24), 6883–6912.

Ross, D.T. (1977). Structured Analysis (SA): A language for communicating ideas, *IEEE Transactions on Software Engineering*, 3, 16–34.

Spicer Patrick Carlo Hector, J. (2007). Integrated reconfigurable cost into the design of multi period scalable reconfigurable manufacturing systems, *The University of Michigan*, 129, 202–212.

8 Actuators and Sensors for Advanced Manufacturing Systems

8.1 INTRODUCTION TO ACTUATORS

An **actuator** is a component of a machine that is responsible for moving and controlling a mechanism or system, for example by opening a valve. In simple terms, it is a "mover".

An actuator requires a control signal and a source of energy. The control signal is relatively low energy and may be electric voltage or current, pneumatic or hydraulic pressure, or even human power. Its main energy source may be an electric current or pneumatic pressure. When it receives a control signal, an actuator responds by converting the source's energy into mechanical motion.

An actuator is a device that converts energy, which may be electric, hydraulic, pneumatic, etc., to mechanical in such a way that it can be controlled. The quantity and the nature of input depend on the kind of energy to be converted and the function of the actuator. Electric and piezoelectric actuators, for instance, work on the input of electric current or voltage; for hydraulic actuators, it is incompressible liquid and for pneumatic actuators, the input is air. The output is always mechanical energy.

Actuators are not something you would read about every day in the media, unlike artificial intelligence and machine learning. But the reality is that they play a critical role in the modern world, almost like no other device ever invented.

8.1.1 COMPONENTS THAT ARE PART OF THE FUNCTIONING OF AN ACTUATOR

- Power source: this provides the energy input that is necessary to drive the actuator. These are often electric or fluid in nature in the industrial sectors.
- Power converter: the role of the power converter is to supply power from the source to the actuator in accordance with the measurements set by the controller. Hydraulic proportional valves and electrical inverters are examples of power converters in industrial systems.
- Actuator: the actual device that converts the supplied energy to mechanical force.
- Mechanical load: the energy converted by the actuator is usually used to make a mechanical device function. The mechanical load refers to this mechanical system that is being driven by the actuator.

DOI: 10.1201/9781003476375-8

- Controller: a controller ensures that the system functions seamlessly with the appropriate input quantities and other setpoints decided by an operator.

8.2 TYPES OF LINEAR ACTUATORS

Depending on the kind of movement they make, and the source of energy used to function, actuators can be classified into different types.

8.2.1 ELECTRIC LINEAR ACTUATOR

As the name implies, electric linear actuators use electrical energy to enable movements in a straight line. They work by moving a piston back and forth based on electrical signals and are mostly used for movements such as pulling, pushing, blocking, lifting, ejecting, clamping or descending.

- Linear actuators function with a motor that generates high-speed rotational motion and a gearbox that slows down its impact. This will, in turn, increase the torque that would be used to turn a lead screw, which results in linear motion of a shaft or drive nut. Often, a 12V DC motor is used in linear actuators, but other voltages can also be used. Switching the polarity of the connection from the motor to the battery would make the motor rotate in the reverse direction.
- Manufacturers offer linear actuators in different strokes, which are implemented by increasing or decreasing the length of the shaft. With different gears, different speeds can also be achieved. Generally speaking, the more the speed of the screw turn, the less the force. A switch within the main actuator shaft at the top and lower end stop the screw as it reaches the end of its movement or stroke. As the shaft reaches its end, the switch cuts off power to the motor.

8.2.2 ELECTRIC ROTARY ACTUATOR

- Electric rotary actuators use electrical energy to achieve rotational movement. This movement can either be continuous or be towards a fixed angle, as seen in servo and stepper motors. Typically speaking, an electric rotary actuator consists of a combination of an electric motor, limit switch and a multiple stage helical gearbox.
- In simple terms, this actuator's operations can be defined like this: when a conductor that carries current is brought within a magnetic field, it will experience a force that is relative to the field's flux density, the current that is flowing through it and its dimensions. Rotation and torque are generated due to the force and back electromotive force (EMF) that ensues.

8.2.3 HYDRAULIC LINEAR ACTUATOR

- The purpose of a hydraulic linear actuator is the same as that of an electric linear actuator–to generate a mechanical movement in a straight line. The difference is that hydraulic linear actuators achieve this with an unbalanced pressure that is applied with hydraulic fluid on a piston in a hollow cylinder that can lead to torque strong enough to move an external object.

• The main advantage of a hydraulic linear actuator is the massive amount of torque it can generate. This is because liquids are almost incompressible. Single-acting hydraulic actuators have pistons that can move in just one direction and a spring is needed for reverse motion. A double-acting hydraulic actuator applies pressure at both ends to facilitate similar movement from both sides.

8.2.4 HYDRAULIC ROTARY ACTUATOR

• Hydraulic rotary actuators make use of incompressible, pressurized fluid to rotate mechanical parts of a device. They mostly come with two kinds of rotational components, circular shafts that have a keyway and tables that have a bolt pattern which can be used to mount other components.

• They are available with single and double shafts. The shaft is rotated when the helical spline teeth on it connect to the corresponding splines on the piston, effectively converting linear movement to rotational motion. When pressure is applied through fluids, the piston moves within the housing, prompting the splines to make the shaft rotate. The shaft can be locked in place when a control valve is shut and fluid is held inside the housing.

8.2.5 PNEUMATIC LINEAR ACTUATOR

• Pneumatic actuators are often considered to be the most cost-effective and simplest of all actuators. Pneumatic linear actuators function using compressed air to create movement, either by extending and retracting a piston or, more rarely, using a carriage that runs on a driveway or a cylindrical tube. The retraction of the piston is either done with a spring or by supplying fluid from the other end.

• Pneumatic linear actuators are best suited to achieve high speed and torque on a relatively small footprint. Quick, point-to-point motion is their strength, and they don't get easily damaged by hard stops. This rugged nature makes them popular in devices that need to be explosion-proof or resistant to difficult conditions like high temperatures.

8.2.6 PNEUMATIC ROTARY ACTUATOR

• Pneumatic rotary actuators use compressed air to produce oscillatory motion. As with pneumatic linear actuators, these are also simple in their design, durable and suitable for work in hazardous environments.

• Three of the most common configurations in pneumatic rotary actuators are rack and pinion, scotch yoke and vane design. In a rack and pinion configuration, the compressed air pushes a piston and rack in linear motion, which in turn causes rotary movements in a pinion gear and output shaft. These can come in single, double or multiple racks.

8.2.7 Piezoelectric Actuators

- Piezo materials are a group of solids like ceramics that react to electrical charge by expanding or contracting and generate energy when a mechanical force is applied. Piezoelectric actuators take advantage of the movement caused by electric signals to create short high-frequency and fast-response strokes. The movement that piezoelectric actuators produce is often parallel to the electric field. However, in some cases, where the device is set to work on the transverse piezoelectric effect, the movement is orthogonal to the electric field.

8.3 CAPABILITIES OF A LINEAR ACTUATOR

Performance metrics are quantifiable outputs that help you evaluate the quality of a particular product. Actuators can be considered under several performance metrics. Traditionally, the most common among them have been torque, speed and durability. These days, energy efficiency is also considered equally important. Other factors that may be considered include volume, mass, operating conditions, etc.

8.3.1 Torque or Force

Naturally, torque is one of the most important aspects to consider in the performance of an actuator. A key factor here is to note that there are two kinds of torque metric to consider: static and dynamic load. Static load torque or force refers to the actuator's capacity when it is at rest. The dynamic metric refers to the device's torque capacity when it is in motion.

8.3.2 Speed

The speed of an actuator differs depending on the weight of the load it is supposed to carry. Usually, the higher the weight, the lower the speed. Hence the speed metric should first be looked at when the actuator is not carrying any load.

8.3.3 Durability

The type of actuator and the manufacturer's design decides the durability of an actuator. Although those such as hydraulic actuators are considered more durable and rugged compared to electric actuators, the detailed specifications on the quality of the material used will be up to the manufacturer.

8.3.4 Energy Efficiency

With increasing concerns about energy conservation and its direct impact on operational costs, energy efficiency is becoming more and more a decisive metric in all kinds of machinery. Here, the lesser the quantity of energy required for an actuator to achieve its goal, the better.

8.4   HOW TO CONNECT LINEAR ACTUATORS

Connecting an electric linear actuator is a rather simple process. Many electric linear actuators come with four pins these days and their connection is as simple as plugging them in. However, if your actuator does not have four pins, the process is slightly different.

8.4.1   PREPARE THE WIRES

Your actuator might come with wires exposed at the end. You can strip back this a bit if required before connecting to a four-pin connector. If the connector's wire is not exposed enough, strip that back as well.

8.4.2   CONNECT THE WIRES

Connect the linear actuator to the four-pin connector by twisting the right exposed wires together and covering them up with electrical tape. Often the wires on the actuator and connector come in blue and brown colors and they can be connected accordingly.

Sometimes, the colors may be different on the actuator. For example, if the actuator has red and black wires, connect the red to the brown wire of the actuator and black to the blue. If it comes with red and blue, connect the red to the brown and blue to the blue wire on the connector. If the wires of the actuator are red and yellow, connect red to the brown wire and yellow to the blue wire.

8.4.3   ALL SET

Now you are good to go. Plug in your connector and plug in the control box to the power socket.

8.5   HOW TO MOUNT A LINEAR ACTUATOR

Choosing an actuator and connecting it properly is only half the job. Equally important is mounting the actuator using a method that is right for your application. Below are two common methods that are used to mount an electric linear actuator.

8.5.1   DUAL PIVOT MOUNTING

This method involves fixing an actuator on both sides with a mounting point that is free to pivot, which usually consists of a mounting pin or a clevis. Dual pivot mounting allows the actuator to pivot on either side as it extends and retracts, allowing the application to achieve a fixed path motion with two free pivot points. One of the most useful applications of this method is to open and close doors. When the actuator extends, the dual fixed points enable the door to swing open. The action of the door closing and opening causes changes in angle, but the pivot provides ample space for the two mounting points to rotate. While using this method, make sure that there is enough room for the actuator to extend, without any obstacles in its way.

8.5.2 STATIONARY MOUNTING

In this method, the actuator is mounted in a stationary position with a shaft mounting bracket fixing it to the shaft. Common uses of this kind of mounting are to achieve action similar to pushing something head-on. For instance, this form of mounting is ideal for switching a button on or off. When deciding on this method, ensure that the mounting apparatus can handle the load of the actuator.

8.6 SENSOR

There are numerous definitions as to what a sensor is, but I would like to define a sensor as an input device which provides an output with respect to a specific physical quantity. The term "input device" in the definition of a sensor means that it is part of a bigger system which provides input to a main control system. Another unique definition of a sensor is as follows: it is a device that converts signals from one energy domain to an electrical domain. The definition of the sensor can be understood if we take an example in to consideration.

The simplest example of a sensor is an LDR or a light dependent resistor. It is a device whose resistance varies according to the intensity of light it is subjected to. When the light falling on an LDR is brighter, its resistance becomes very low and when the light is low, well, the resistance of the LDR becomes very high.

8.7 CLASSIFICATION OF SENSORS

There are several classifications of sensors used by different authors and experts. Some are very simple and some are very complex. The following classification of sensors may already be used by an expert in the subject, but this is a very simple classification of sensors.

8.7.1 ACTIVE SENSORS

These are those which require an external excitation signal or a power signal.

8.7.2 PASSIVE SENSORS

These do not require any external power signal and directly generate output response.

The other type of classification is based on the means of detection used in the sensor. Some of the means of detection are electric, biological and chemical, radioactive, etc. The next classification is based on conversion phenomenon, i.e., the input and the output. Some of the common conversion phenomena are photoelectric, thermoelectric, electrochemical, electromagnetic, etc.

The final classification of sensors are analog and digital sensors. Analog sensors produce an analog output, i.e., a continuous output signal with respect to the quantity being measured. Digital sensors, in contrast to analog sensors, work with discrete or digital data. The data in digital sensors, which is used for conversion and transmission, is digital in nature.

8.8   DIFFERENT TYPES OF SENSORS

The following is a list of different types of sensors that are commonly used in various applications. All these sensors are used for measuring one of the physical properties like temperature, resistance, capacitance, conduction, heat transfer, etc.

- Temperature sensor;
- Proximity sensor;
- Accelerometer;
- IR sensor (infrared sensor);
- Pressure sensor;
- Light sensor;
- Ultrasonic sensor;
- Touch sensor;
- Color sensor;
- Humidity sensor;
- Flow and level sensor.

8.8.1   TEMPERATURE SENSOR

One of the most common and most popular sensors is the temperature sensor. A temperature sensor, as the name suggests, senses the temperature, i.e., it measures the changes in the temperature. In a temperature sensor, the changes in the temperature correspond to change in its physical property, such as resistance or voltage.

There are different types of temperature sensors like temperature sensor ICs (like LM35), thermistors, thermocouples, resistive temperature devices (RTD), etc., temperature sensors are used everywhere in computers, mobile phones, automobiles, air conditioning systems, industries, etc.

8.8.2   PROXIMITY SENSORS

A proximity sensor is a non-contact type of sensor that detects the presence of an object. Proximity sensors can be implemented using different techniques, such as optical (like infrared or laser), ultrasonic, hall effect, capacitive, etc. Some of the applications of proximity sensors are in mobile phones, cars (parking sensors), industries (object alignment), ground proximity in aircraft, etc.

8.8.3   INFRARED SENSOR (IR SENSOR)

IR sensors or infrared sensors are light based sensors that are used in various applications like proximity and object detection. IR sensors are used as proximity sensors in almost all mobile phones. There are two types of infrared or IR sensors: transmissive type and reflective type.

In a transmissive type of IR sensor, the IR transmitter (usually an IR LED) and the IR detector (usually a photo diode) are positioned facing each other so that when an object passes between them, the sensor detects the object. The other type of IR sensor is a reflective type of IR sensor. In this, the transmitter and the detector are positioned adjacent to each other facing the object. When an object comes in front of the sensor, the sensor detects the object. Different applications where an IR sensor is implemented are mobile phones, robots, industrial assembly, automobiles, etc.

8.8.4 ULTRASONIC SENSOR

An ultrasonic sensor is a non-contact type device that can be used to measure distance as well as the velocity of an object. An ultrasonic sensor works based on the properties of the sound waves with a frequency greater than that of the human audible range. Using the time of flight of the sound wave, an ultrasonic sensor can measure the distance of the object. The Doppler Shift property of the sound wave is used to measure the velocity of an object.

THEORETICAL QUESTIONS

Question 1. What is an actuator? Explain the function of actuator with neat and clean diagrams.

Question 2. Explain the working of different types of actuators with neat and clean diagrams.

Question 3. Explain the principle of an actuator and also explain its importance.

Question 4. Discuss the components of an actuator.

Question 5. Define the principle of a sensor and its importance.

Question 6. Explain the working of different types of sensors with neat and clean diagrams.

Question 7. Explain the merits and demerits of actuators.

Question 8. Explain the advantages and disadvantages of sensors.

Question 9. Explain the components of sensors.

Question 10. Differentiate between actuators and sensors with the help of suitable examples.

MULTIPLE CHOICE QUESTIONS

Question 1. A pressure control process using proportional plus integral control has a time constant of ten seconds. The best choice of actuator would be:
a) An electric motor;
b) A pneumatic diaphragm;
c) A piston and cylinder;
d) A solenoid electrical.

Question 2. An electronic controller creates a 4 to 20 mA dc signal that must actuate a steam valve for temperature control. The best and most economical choice would be to:
a) Use an all-electric actuator system;
b) Convert to a pneumatic signal at the controller and use a pneumatic actuator;

    c) Use a pneumatic actuator with an electric to pneumatic valve positioner;
    d) None of the above.

Question 3. The basic function of the spring in a control valve is to:
    a) Characterize flow;
    b) Oppose the diaphragm so as to position the valve according to signal pressure;
    c) Close the valve if air failure occurs;
    d) Open the valve if air failure occurs.

Question 4. A single seated globe valve containing a plug 1½ inches in diameter is used in a line pressurized to 500 psi. What actuator force is required for tight shutoff?
    a) 884 pounds;
    b) 2,000 pounds;
    c) Depends upon direction of flow through the valve;
    d) None of the above.

Question 5. A diaphragm actuator has a diaphragm area of 50 square inches and is adjusted to stroke a valve when a 3 to 15 psi (20 to 100 k Pa) signal is applied. If the signal is 15 psi (100 k pa) the force on the valve stem will be:
    a) 750 pounds;
    b) 750 pounds less than the opposing spring force;
    c) Dependent on hysteresis;
    d) None of the above.

Question 6. Assume that a control valve regulates stream flow to a process and that high temperature makes the reaction hazardous. The usual pneumatically operated control valve utilizes the following action for fail-safe operation:
    a) Air to open;
    b) Air to close;
    c) 3 psi (20 kpa) to fully open;
    d) 15 psi (100 kpa) to fully close.

Question 7. One advantage of an electric to pneumatic valve positioner is:
    a) It can be used on flow control;
    b) It produces positive valve position;
    c) It conserves energy;
    d) It dampers valve travel.

Question 8. A valve positioner:
    a) Takes the place of a cascade control system;
    b) Provides more precise valve position;
    c) Makes a pneumatic controller unnecessary;
    d) Provides a remote indication of valve position.

Question 9. A diaphragm actuator has a diaphragm area of 115 square inches. A valve positioner is attached to the actuator and fed with 22 psi air supply. If after 9 psi a signal is received from the controller the signal changes to 10 psi and the valve fails to move. What is the force applied to the valve stem?
    a) 2.530 pounds;
    b) 1.495 pounds;

    c) 1.035 pounds;
    d) None of the above.

Question 10. A high-pressure flow process requires a valve with tight packing. This would suggest that:
    a) A valve positioner should be employed;
    b) The actuator must be sized to provide adequate force;
    c) Oversized pneumatic signals lines are required;
    d) The controller supplying the signal to the valve must have a very narrow proportional band.

Question 11. Actuators are used to:
    a) Sense an object;
    b) Activate a chemical;
    c) Make a mechanical movement;
    d) All of the above.

Question 12. What is the formula of speed control valve during extension of a flow control valve?
    a) $V = (Q/A)$;
    b) $V = Q.A$;
    c) $V = A/Q$;
    d) $V = Q(A - a)$.

Question 13. Which among the following are not the main selection criteria of the control valves?
    a) Type of actuation;
    b) Environmental conditions;
    c) Space requirement;
    d) Software support.

Question 14. The valve packing of control valves is used:
    a) To prevent the fluid from escaping;
    b) To control the force generated by actuators;
    c) To control different parameters of the fluid;
    d) To control the direction of flow.

Question 15. Which among the following fluid parameters are not controlled by the control valves?
    a) Pressure;
    b) Rate of flow;
    c) Speed;
    d) Direction of flow.

Question 16. What is the formula of speed control valve during retraction of a flow control valve?
    a) $V = (Q/A)$;
    b) $V = Q.A$;
    c) $V = A/Q$;
    d) $V = Q/(A - a)$.

Question 17. What is the function of the pressure control valve?
a) To control the force generated by actuators;
b) To perform two operations in sequence;
c) To control the direction of flow;
d) To avoid the development of excess of pressure.

Question 18. What is the function of the control valve?
a) To control different parameters of the fluid;
b) To perform two operations in sequence;
c) To control the direction of flow;
d) To avoid the development of excess of pressure.

Question 19. Which among the following are not the "work parameters" of the fluid?
a) Direction;
b) Speed;
c) Pressure;
d) Temperature of flow.

Question 20. There is no difference between the control valve of the pneumatic and hydraulic system.
a) True;
b) False.

Question 21. The pressure relief valve is an important component which is required for every positive displacement pump.
a) True;
b) False.

Question 22. Which of the following is correct for tactile sensors?
a) Touch sensitive;
b) Pressure sensitive;
c) Input voltage sensitive;
d) Humidity sensitive.

Question 23. Change in output of sensor with change in input is:
a) Threshold;
b) Slew rate;
c) Sensitivity;
d) None of the above.

Question 24. Which of the following can be cause for non-zero output when zero input?
a) Bias;
b) Slew;
c) Offset;
d) Offset or bias.

Question 25. Sensitivity of a sensor can be depicted by:
a) Niquist plot;
b) Pole–zero plot;
c) Bode plot;
d) None of the above.

Question 26. Which of the following errors is caused by a reversal of measured property?
   a) Hysteresis;
   b) Noise;
   c) Digitization error;
   d) Quantization error.

Question 27. The smallest change which a sensor can detect is:
   a) Resolution;
   b) Accuracy;
   c) Precision;
   d) Scale.

Question 28. Thermocouples generate output voltage according to:
   a) Circuit parameters;
   b) Humidity;
   c) Temperature;
   d) Voltage.

Question 29. A sensor is a type of transducer.
   a) True;
   b) False.

Question 30. Which of the following is not an analog sensor?
   a) Potentiometer;
   b) Force-sensing resistors;
   c) Accelerometers;
   d) None of the above.

ANSWERS TO MCQs

1 a; 2 c; 3 b; 4 c; 5 b; 6 a; 7b; 8 b; 9 a; 10 b 11 c; 12 a; 13 d; 14 a; 15 c; 16 d; 17 a; 18 a; 19 d; 20 b; 21 a; 22 a; 23 c; 24 d; 25 c; 26 a; 27 a; 28 c; 29 a; 30 d.

BIBLIOGRAPHY

Abdi, M.R. (2009). Fuzzy multi criteria decision model for evaluating reconfigurable machines, *International Journal of Production Economics*, 117, 1–15.

Bi, Z.M., & Wang, L. (2009). Optimal design of reconfigurable parallel machining system, *Journals of Robotics and Computer Integrated Manufacturing*, 25, 951–961.

Carlos, A. (2009). Coello Coello Ricardo L and Becerra evolutionary multiobjective optimization in materials science and engineering, *Journals of Material and Manufacturing Processes*, 24, 119–129.

Ferguson, S., Siddiqi, A., Lewis, K., & Weck, O. (2007). Flexible and reconfigurable systems: Nomenclature and review, *Proceeding of the ASME international design engineering technical conference and computer and information in engineering conference*, September 4–7, 1–13.

Maraghy, R., & Ulay, Y. (2007). Reconfigurable manufacturing systems: Key to future manufacturing, *Journal of Intelligent Manufacturing*, 11(4), 403–419.

Nambiar, N. (2010). Modern manufacturing Paradigm—A comparision, *Proceeding of the international multi conference of engineers and computer scientists*, March 17–19, 5–8.

Rao, R.V., & Padmanabhan, K.K. (2006). Selection, identification and comparison of industrial robots using digraph and matrix methods, *Robotics and Computer-Integrated Manufacturing*, 22(4), 373–383.

Song, S., Li, A., & Xu, L. (2007). Study of CAPP system suited for reconfigurable manufacturing system, *Journal of Advanced Manufacturing Systems*, 2, 5769–5772.

9 Robotic Technology

9.1 INTRODUCTION TO ROBOTS

Robotics is a branch of engineering and science that includes electronics engineering, mechanical engineering, computer science and so on. This branch deals with the design, construction, use of control of robots, sensory feedback and information processing. These are some technologies which will replace humans and human activities in coming years. These robots are designed to be used for any purpose, but these are used in sensitive environments such as bomb detection, deactivation of various bombs, etc. Robots can take any form but many of them have been given a human appearance. The robots with a human appearance may have to walk like humans, and have speech, cognition, and most importantly all the things a human can do. Most of the robots of today are inspired by nature and are known as bio-inspired robots.

Robotics is that branch of engineering that deals with conception, design, operation and manufacturing of robots. Three laws state that:

- Robots will never harm human beings.
- Robots will follow instructions given by humans without breaking law one.
- Robots will protect themselves without breaking other rules.

9.2 CHARACTERISTICS

Some characteristics of robots are given below.

9.2.1 APPEARANCE

Robots have a physical body. They are held by the structure of their body and are moved by their mechanical parts. Without appearance, robots would be just a software program.

DOI: 10.1201/9781003476375-9

9.2.2 Brain

Another name for the brain in robots is on-board control unit. Using this the robot receives information and sends commands as output. With this control unit the robot knows what to do otherwise it will be just a remote-controlled machine.

9.2.3 Sensors

The use of these sensors in robots is to gather information from the outside world and send it to the brain. Basically, these sensors have circuits in them that produce a voltage in them.

9.2.4 Actuators

The robots move, and the parts which help these robots move are called actuators. Some examples of actuators are motors, pumps and compressors, etc. The brain tells these actuators when and how to respond or move.

9.2.5 Program

Robots only work or respond to the instructions which are provided to them in the form of a program. These programs only tell the brain when to perform which operation, for example: when to move, produce sounds, etc. These programs only tell the robot how to use the sensors data to make decisions.

9.2.6 Behavior

A robot's behavior is decided by the program which has been built for it. Once the robot starts making the movement, one can easily tell which kind of program is being installed inside the robot.

9.3 TYPES OF ROBOTS

9.3.1 Articulated

The feature of this robot is its rotary joints and these range from two to ten or more joints. The arm is connected to the rotary joint and each joint, known as the axis, provides a range of movements.

9.3.2 Cartesian

These are also known as gantry robots. These have three joints which use the cartesian coordinate system, i.e., x, y, z. These robots are provided with attached wrists to provide rotatory motion.

9.3.3 CYLINDRICAL

These types of robots have at least one rotatory joint and one prismatic joint which are used to connect the links. The use of rotatory joints is to rotate along the axis and the prismatic joint is used to provide linear motion.

9.3.4 POLAR

These are also known as spherical robots. The arm is connected to the base with a twisting joint and has a combination of two rotatory joints and one linear joint.

9.3.5 SCARA

These robots are mainly used in assembly applications. The arm is cylindrical in design. It has two parallel joints which are used to provide compliance in one selected plane.

9.3.6 DELTA

The structure of these robots is spider-shaped. They are built by joint parallelograms that are connected to a common base. The parallelogram moves in a dome-shaped work area. These are mainly used in food and electrical industries.

9.4 SCOPE AND LIMITATIONS OF ROBOTS

The advanced versions of machines are robots which are used to do advanced tasks and are programmed to make decisions on their own. When a robot is designed, the most important thing to be kept in mind is that what function is to be performed and what are the limitations of the robot. Each robot has a basic level of complexity and each of the levels has a scope which limits the functions that can be performed. For general basic robots, their complexity is decided by the number of limbs, actuators and sensors that are used, while for advanced robots the complexity is decided by the number of microprocessors and microcontrollers used. Increasing any component in the robot increases the scope of the robot.

9.4.1 ADVANTAGES

The advantages of using robots are:

- They can get information that a human can't get.
- They can perform tasks without any mistakes, and very efficiently and quickly.
- Maximum robots are automatic, so they can perform different tasks without needing human interaction.
- Robots are used in different factories to produce items like planes, car parts, etc.
- They can be used for mining purposes.

9.4.2 Disadvantages

The disadvantages of using robots are:

- They need a power supply to keep going. People working in factories may lose their jobs as robots can replace them.
- They need high maintenance to keep them working all day long. The cost of maintaining the robots can be expensive.
- They can store huge amounts of data, but they are not as efficient as our human brains.
- As we know, robots work on the program that has been installed in them. Other than the program installed, robots can't do anything different.
- The most important disadvantage is that if the program of robots is in the wrong hands they can cause a huge amount of destruction.

9.4.3 Applications

Different types of robots can perform different types of tasks. For example, many of the robots are made for assembly work, which means that they are not relevant for any other work and these types of robots are called assembly robots. Similarly, for seam welding many suppliers provide robots with their welding materials and these types of robots are known as welding robots. On the other hand, many robots are designed for heavy-duty work and are known as heavy duty robots.

Some applications are:

- A robot can also do herding tasks.
- Robots are increasingly used more than humans in manufacturing, while in the auto-industry more than half of the labors are performed by "robots".
- Many of the robots are used as military robots.
- Robots have been used in cleaning up areas such as toxic waste or industrial wastes, etc.
- Agricultural robots.
- Household robots.
- Domestic robots.

THEORETICAL QUESTIONS

Question 1. Discuss the function and applications of robots.

Question 2. Explain the construction, working and principle of robots with neat and clean diagrams, along with their merits and demerits.

Question 3. Explain the components of robots with suitable applications.

Question 4. Explain a robotic control system with suitable examples.

Question 5. Discuss the concept of sensor technology with suitable examples.

Question 6. Describe industrial robotic applications.

Question 7. Explain robotic motion with suitable examples.

MULTIPLE CHOICE QUESTIONS

Question 1. What is the name for information sent from robot sensors to robot controllers?
 a) Temperature;
 b) Pressure;
 c) Feedback;
 d) Signal.

Question 2. Which of the following terms refers to the rotational motion of a robot arm?
 a) Swivel;
 b) Axle;
 c) Retrograde;
 d) Roll.

Question 3. What is the name for the space inside which a robot unit operates?
 a) Environment;
 b) Spatial base;
 c) Work envelope;
 d) Exclusion zone.

Question 4. Which of the following terms IS NOT one of the five basic parts of a robot?
 a) Peripheral tools;
 b) End effectors;
 c) Controller;
 d) Drive.

Question 5. Decision support programs are designed to help managers make:
 a) Budget projections;
 b) Visual presentations;
 c) Business decisions;
 d) Vacation schedules.

Question 6. The number of moveable joints in the base, the arm and the end effectors of the robot determines:
 a) Degrees of freedom;
 b) Payload capacity;
 c) Operational limits;
 d) Flexibility.

Question 7. Which of the following places would be LEAST likely to include operational robots?
 a) Warehouse;
 b) Factory;
 c) Hospitals;
 d) Private homes.

Question 8. For a robot unit to be considered a functional industrial robot, typically, how many degrees of freedom would the robot have?

    a)   Three;
    b)   Four;
    c)   Six;
    d)   Eight.

Question 9. Which of the following terms refers to the use of compressed gasses to drive (power) the robot device?

    a)   Pneumatic;
    b)   Hydraulic;
    c)   Piezoelectric;
    d)   Photosensitive.

Question 10. With regard to the physics of power systems used operate robots, which statement or statements are most correct?

    a)   Hydraulics involves the compression of liquids;
    b)   Hydraulics involves the compression of air;
    c)   Pneumatics involves the compression of air;
    d)   Chemical batteries produce AC power.

Question 11. The original LISP machines produced by both LMI and Symbolics were based on research performed at:

    a)   CMU;
    b)   MIT;
    c)   Stanford University;
    d)   RAMD.

Question 12. Which of the following statements concerning the implementation of robotic systems is correct?

    a)   Implementation of robots CAN save existing jobs;
    b)   Implementation of robots CAN create new jobs;
    c)   Robotics could prevent a business from closing;
    d)   All of the above.

Question 13. Which of the following IS NOT one of the advantages associated with a robotics implementation program?

    a)   Low costs for hardware and software;
    b)   Robots work continuously around the clock;
    c)   Quality of manufactured goods can be improved;
    d)   Reduced company cost for worker fringe benefits.

Question 14. Which of the following "laws" is Asimov's first and most important law of robotics?

    a)   Robot actions must never result in damage to the robot;
    b)   Robots must never take actions harmful to humans;
    c)   Robots must follow the directions given by humans;
    d)   Robots must make greater profits for businesses.

Question 15. In a rule-based system, procedural domain knowledge is in the form of:

    a)   Production rules;
    b)   Rule interpreters;

    c)   Meta-rules;
    d)   Control rules.

Question 16. If a robot can alter its own trajectory in response to external conditions, it is considered to be:
    a)   Intelligent;
    b)   Mobile;
    c)   Open loop;
    d)   Non-servo.

Question 17. One of the leading American robotics centers is the Robotics Institute located at:
    a)   CMU;
    b)   MIT;
    c)   RAND;
    d)   SRI.

Question 18. A mechanism having its motive power so concealed that it appears to move spontaneously is called:
    a)   Automatic;
    b)   Clock Jack;
    c)   Robot;
    d)   Automata.

Question 19. Which of the following is not an advantage of robots?
    a)   They can assist humans with disabilities;
    b)   They can replace jobs;
    c)   They can be used in dangerous environments;
    d)   They don't get tired or require a break.

Question 20. The branch of technology that deals with the design, construction, operation and application of robots is:
    a)   Levers;
    b)   Robotics;
    c)   Creative power;
    d)   Science CSF.

Question 21. When working in a group for robotics, students should:
    a)   Stay on task but don't work with other group members;
    b)   Socialize with group members outside of your group and then work alone;
    c)   Socialize with other group members and don't help your group;
    d)   Stay on task and work with other group members appropriately.

Question 22. Isaac Asimov first announced the three laws of robotics in 1942. Which is the third law:
    a)   A robot may not injure a human being or, through inaction, allow a human being to come to harm;
    b)   A robot can't go to school;
    c)   A robot must obey orders given to it by human beings except where such orders would conflict with the first law;
    d)   A robot must protect its own existence as long as such protection does not conflict with the first or second law.

Question 23. How many systems does a robot have?
- a) Two;
- b) Six;
- c) Four;
- d) Three.

Question 24. How many types of robots are there?
- a) Seven;
- b) Ten;
- c) Six;
- d) Eight.

Question 25. Which one of these is NOT a type of robot:
- a) Medical;
- b) Industrial;
- c) Household;
- d) Apologetic.

Question 26. The small mobile robot base is used in the Robot Educator. This robot is able to perform some but not all of the tasks in robotics engineering activities. Which are these:
- a) Light sensor;
- b) Lego mindstorms education software;
- c) Robot;
- d) Robot educator model (REM).

Question 27. A machine that is able to interact with and respond to its environment is characterized by three central capabilities: the ability to sense, the ability to plan and the ability to enact:
- a) Code;
- b) Taskbot;
- c) Robots;
- d) Ports.

Question 28. The three characteristic capabilities that define a robot are:
- a) Comment;
- b) Sensor;
- c) Sense-plan-act;
- d) NXT brick.

Question 29. When working in a group for robotics, students should:
- a) Socialize instead of work and then work alone;
- b) Stay on task and don't work with your group;
- c) Work alone and don't socialize with group members;
- d) Stay on task and work with group members appropriately.

Question 30. What is a general term for any command or group of commands in a program. In the NXT Programming Software, which is one or more blocks:
- a) Comment;
- b) Code;
- c) Ports;
- d) Robot.

ANSWERS TO MCQs

1 c; 2 d; 3 c; 4 a; 5 c; 6 a; 7 d; 8 c; 9 a; 10 c; 11 b; 12 b; 13 d; 14 a; 15 b; 16 a; 17 a; 18 a; 19 d; 20 b; 21 b; 22 d; 23 c; 24 d; 25 d; 26 d; 27 c; 28 c; 29 d; 30 b.

BIBLIOGRAPHY

Abdi, M.R., & Labib, A.W. (2003). A design strategy for reconfigurable manufacturing systems (RMSs) using analytical hierarchical process (AHP): A case study, *International Journal of Production Research*, 41(10), 2273–2299.

Agarwal, A., & Sarkis, J. (1998). A review and analysis of comparative performance studies on functional and cellular manufacturing layouts, *Computers & Industrial Engineering*, 1, 77–89.

Albayrakoglu, M.M. (1996). Justification of new manufacturing technology: A strategical approach using the Analytical Hierarchy Process, *Production and Inventory Management Journal, First Quarter*, 1, 71–76.

Arbel, A., & Seidmann, A. (1984). Performance evaluation of flexible manufacturing systems, *IEEE Transactions on Systems, Man and Cybernetics*, 4, 606–617.

Ismail, N., Musharavati, F., Hamouda, A.S.M., & Ramil, A.R. (2008). Manufacturing process planning optimization in reconfigurable multiple parts flow lines, *Journal of Achievements in Material and Manufacturing Engineering*, 31(2), 671–677.

Rao, Venkata., & Padmanabhan, K. K. (2007). Rapid prototyping process selection using graph theory and matrix approach, *Journal of Material and Processing Technology*, 194, 81–88.

10 Material Handling for AMS

10.1 INTRODUCTION TO MATERIAL HANDLING

Material handling is a necessary and significant component of any productive activity. It is something that goes on in every plant all the time. Material handling means providing the right amount of the right material, in the right condition, at the right place, at the right time, in the right position and for the right cost, by using the right method. It is simply picking up, moving and laying down of materials through manufacture. It applies to the movement of raw materials, parts in process, finished goods, packing materials and disposal of scrap. In general, hundreds and thousands of tons of materials are handled daily, requiring the use of large amounts of manpower while the movement of materials takes place from one processing area to another, or from one department to another department of the plant. The cost of material handling contributes significantly to the total cost of manufacturing. Material handling is the field concerned with solving the pragmatic problems involving the movement, storage, control and protection of materials, goods and products throughout the processes of cleaning, preparation, manufacturing, distribution, consumption and disposal of all related materials, goods and their packaging. The focus of studies of material handling course work is on the methods, mechanical equipment, systems and related controls used to achieve these functions. The material handling industry manufactures and distributes the equipment and services required to implement material handling systems, from obtaining, locally processing and shipping raw materials to utilization of industrial feedstocks in industrial manufacturing processes.

Material handling systems range from a simple pallet rack and shelving projects, to complex conveyor belt and automated storage and retrieval systems (AS/RS); from mining and drilling equipment to custom built barley malt drying rooms in breweries. Material handling can also consist of sorting and picking, as well as automatic guided vehicles. In the modern era of competition, this has acquired greater importance due to the growing need for reducing manufacturing costs. The importance of the material handling function is greater in those industries where the ratio of handling cost to processing cost is large. Today material handling is rightly considered as one of the most potentially lucrative areas for reduction of costs. A properly designed and integrated

DOI: 10.1201/9781003476375-10

material handling system provides tremendous cost saving opportunities and customer services improvement potential.

10.2   PRINCIPLES OF MATERIAL HANDLING

1. Planning principle: all handling activities should be planned.
2. Systems principle: plan a system integrating as many handling activities as possible and coordinating the full scope of operations (receiving, storage, production, inspection, packing, warehousing, supply and transportation).
3. Space utilization principle: make optimum use of cubic space.
4. Unit load principle: increase quantity, size and weight of load handled.
5. Gravity principle: utilize gravity to move a material wherever practicable.
6. Material flow principle: plan an operation sequence and equipment arrangement to optimize material flow.
7. Simplification principle: reduce, combine, or eliminate unnecessary movement and/or equipment.
8. Safety principle: provide for safe handling methods and equipment.
9. Mechanization principle: use mechanical or automated material handling equipment.
10. Standardization principle: standardize method, types and size of material handling equipment.
11. Flexibility principle: use methods and equipment that can perform a variety of tasks and applications.
12. Equipment selection principle: consider all aspects of materials, moves and methods to be utilized.
13. Dead weight principle: reduce the ratio of dead weight to pay load in mobile equipment.
14. Motion principle: equipment designed to transport material should be kept in motion.
15. Idle time principle: reduce idle time/unproductive time of both material handling equipment and manpower.
16. Maintenance principle: plan for preventive maintenance or scheduled repair of all handling equipment.
17. Obsolescence principle: replace obsolete handling methods/equipment when more efficient methods/equipment will improve operation.
18. Capacity principle: use handling equipment so as to help achieve its full capacity.
19. Control principle: use material handling equipment to improve production control, inventory control and other handling.
20. Performance principle: determine efficiency of handling performance in terms of cost per unit handled, which is the primary criteria.

10.3   TYPES OF MATERIAL HANDLING EQUIPMENT

The four main categories of material handling equipment include storage, engineered systems, industrial trucks and bulk material handling.

10.3.1 Storage and Handling Equipment

Storage equipment is used to hold or buffer materials during times when they are not being transported. These periods could refer to temporary pauses during long-term transportation or long-term storage designed to allow the buildup of stock. The majority of storage equipment refers to pallets, shelves or racks onto which materials may be stacked in an orderly manner to await transportation or consumption. Many companies have investigated increased efficiency possibilities in storage equipment by designing proprietary packaging that allows materials or products of a certain type to conserve space while in inventory.

Examples of storage and handling equipment include:

- Racks, such as pallet racks, drive-through or drive-in racks, push-back racks and sliding racks, are a basic but important method of storage, saving floor space while keeping their contents accessible.
- Stacking frames are stackable like blocks, as their name implies. They allow crushable pallets of inventory, such as containers of liquid, to be stacked to save space without damage.
- Shelves, another basic storage method, are less open than racks. Used with bins and drawers, they're more able to keep smaller and more difficult to manage materials and products stored and organized. Shelving types can include boltless, cantilever, revolving and tie-down.
- Mezzanines, a type of indoor platform, help to create more floor space in a warehouse or other storage building for offices or more storage. Typical types include modular, movable, rack supported, building supported and free-standing versions.
- Work assist tooling enables safe and efficient product handling across numerous industries in applications that require the movement of products, enhancing the efficiency of assembly and manufacturing operations.

10.3.2 Engineered Systems

Engineered systems cover a variety of units that work cohesively to enable storage and transportation. They are often automated. A good example of an engineered system is an automated storage and retrieval system, often abbreviated to AS/RS, which is a large automated organizational structure involving racks, aisles and shelves accessible by a "shuttle" system of retrieval. The shuttle system is a mechanized cherry picker that can be used by a worker or can perform fully automated functions to quickly locate a storage item's location and retrieve it for other uses. Other types of engineered systems include:

- Conveyor systems come in a variety of types, depending on what they are meant to transport, including vibrating, overhead, chain, vertical and apron conveyors.
- Automatic guided vehicles are independent computer-operated trucks that transport loads along a predetermined path, with sensors and detectors to avoid bumping into anything.

10.3.3 INDUSTRIAL MATERIAL HANDLING TRUCKS

Industrial trucks refer to the different kinds of transportation items and vehicles used to move materials and products in materials handling. These transportation devices can include small hand-operated trucks, pallet jacks and various kinds of forklifts. These trucks have a variety of characteristics to make them suitable for different operations. Some trucks have forks, as in a forklift, or a flat surface with which to lift items, while some trucks require a separate piece of equipment for loading. Trucks can also be manual or powered lift and operation can be walk or ride, requiring a user to manually push them or to ride along on the truck. A stack truck can be used to stack items, while a non-stack truck is typically used for transportation and not for loading.

There are many types of industrial trucks:

- Hand trucks, one of the most basic pieces of material handling equipment, feature a small platform to set the edge of a heavy object on and a long handle to use for leverage. Whatever is being moved must be tipped so that it rests on the handle, and is carried at a tilt to its destination.
- Pallet trucks, also known as pallet jacks, are a type of truck specifically for pallets. They slide into a pallet and lift it up to move it. Pallet trucks come in both manual and electric types.
- Platform trucks are hand trucks low to the ground, with a wide platform for transporting goods.
- Side loaders, also known as VNA (very narrow aisle) trucks, are meant to fit in narrow warehouse aisles, as they can load objects from different directions. They're also good for long, awkward products that need moving. Automatic guided vehicles, as discussed above, shuttle products along a route automatically, without human guidance.

10.3.4 BULK MATERIAL HANDLING EQUIPMENT

Bulk material handling refers to the storing, transportation and control of materials in loose bulk form. These materials can include food, liquid or minerals, among others. Generally, these pieces of equipment deal with the items in loose form, such as conveyor belts or elevators designed to move large quantities of material, or in packaged form, through the use of drums and hoppers.

- Conveyors, as mentioned above, come in a wide variety of types for different types of bulk material.
- Stackers, which are usually automated, pile bulk material onto stockpiles, moving between two points along rails in a yard.
- Bucket elevators, also known as grain legs, use buckets attached to a rotating chain or belt to carry material vertically.
- Grain elevators are tall buildings specifically for storing grain. They include equipment to convey the grain to the top of the elevator, where it is sent out for processing.

- Hoppers are funnel-shaped containers that allow material to be poured or dumped from one container to another. Unlike a funnel, though, hoppers can hold material until it's needed and then release it.

10.3.4.1 Advantages of Implementing Automated Material Handling Systems

- Save money.

Automated systems speed up productivity. The work can be done faster and with fewer people than with manual equipment. Business owners save money by hiring fewer employees. Automated material handling systems are an economical investment.

- Improve efficiency.

Workers are better able to do their tasks using automated equipment, meaning they can work much more efficiently. It takes less time to do a job and it is more likely to be done correctly, which means each employee is able to accomplish more.

- Maximize space.

Automated systems like lifts make it easier to store materials and products. Items can be stacked higher and be accessed more easily, opening up the workspace.

- Reduce accidents.

One of the biggest advantages to automated equipment is that fewer accidents occur. Automated equipment can eliminate the need for heavy lifting, reduce the chances of workers tripping or falling and make workstations ergonomic to prevent injury.

- Better customer service.

Make customers happy by fulfilling orders faster, reducing or eliminating mistakes and improving shipping. You can even take on new customers because increased productivity allows you to fulfill more orders.

- Increase warehouse value.

The resale value of your warehouse and equipment goes up with automated systems. In the event that you sell your workspace, having automated systems in place not only attracts more buyers, but it means you can ask a higher price for the entire package or the individual pieces of equipment.

- Attract employees.

If you're looking for good quality workers in your warehouse or plant, having automated systems in place will help to recruit a higher caliber of employees.

Workers with experience and more specific training to handle automated systems will be interested in working for your company.

10.3.4.2 Disadvantages of Implementing Automated Material Handling Systems

* Initial cost of equipment.

Automated equipment is more expensive up front than manual equipment. However, what you save in manpower and what you gain in increased productivity means the equipment will eventually pay for itself and then some.

* Reduced flexibility for change.

Once automated systems are in place, it is likely not as easy to make changes in your workspace. But once you go automated and see how smoothly everything runs, it's not likely you'll want to return to manual equipment afterward.

* Possible downtime due to malfunction.

With automatic machines there is always the chance of a problem or breakdown, which can lead to considerable downtime while it is repaired. If the problem cannot be fixed by anyone on site, an outside specialist may need to be called, which could mean more time spent waiting. In some cases, work may be able to continue manually in the meantime, but in some cases that isn't possible. Equipment malfunction can be avoided in most cases with routine maintenance of all machinery. If proper care is taken to keep automated systems in good shape, breakdowns should only occur rarely, if ever.

* Maintenance costs.

Some automated equipment needs maintenance. Routine maintenance may be performed regularly by onsite workers. Properly maintained equipment will save you money in the long run by preventing problems and increasing the overall lifespan of your systems. Carolina Material Handling offers many types of automated equipment that require very low maintenance, keeping costs low.

THEORETICAL QUESTIONS

Question 1. Define a material handling system with suitable examples.

Question 2. Describe the principle of material handling for AMS with applications.

Question 3. Explain the types of material handling for AMS with neat diagrams.

Question 4. Describe the components of material handling for AMS with neat diagrams.

Question 5. Discuss merits and demerits of material handling systems.

Question 6. Describe the applications of material handling systems related to manufacturing industry.

MULTIPLE CHOICE QUESTIONS

Question 1. Material handling consists of movement of material from:
   a)   One machine to another;
   b)   One shop to another shop;
   c)   Stores to shop;
   d)   All of the above.

Question 2. Economy in material handling can be achieved by:
   a)   Employing gravity feed movements;
   b)   Minimizing distance of travel;
   c)   By carrying material to destination without using manual labor;
   d)   All of the above.

Question 3. The principle of "unit load" states that:
   a)   Materials should be moved in lots;
   b)   One unit should be moved at a time;
   c)   Both (a) and (b);
   d)   None of the above.

Question 4. A forklift truck is used for:
   a)   Lifting and lowering;
   b)   Vertical transportation;
   c)   Both (a) and (b);
   d)   None of the above.

Question 5. A wheelbarrow is used for:
   a)   Lifting and lowering;
   b)   Vertical transportation;
   c)   Both (a) and (b);
   d)   None of the above.

Question 6. Cranes are used for:
   a)   Lifting and lowering;
   b)   Vertical transportation;
   c)   Both (a) and (b);
   d)   None of the above.

Question 7. An overbridge crane has:
   a)   Transverse movement;
   b)   Longitudinal movement;
   c)   Both (a) and (b);
   d)   None of the above.

Question 8. The following is used to transport material with a flat bottom:
   a)   Belt conveyer;
   b)   Roller conveyor;
   c)   Chain conveyor;
   d)   None of above.

Question 9. Special purpose material handling equipment is used in:
   a)   Process layout;
   b)   Line layout;
   c)   Both (a) and (b);
   d)   None of the above.

Question 10. Which of the following applications is a belt conveyor used for?
 a)   Material transportation over a long distance;
 b)   Material transportation within premises;
 c)   Material transportation for processing;
 d)   All of the above.

Question 11. Slat belts are made up of:
 a)   Wood;
 b)   Plastic;
 c)   Metal;
 d)   Any of the above.

Question 12. Pneumatic conveying is done under which of the following conditions?
 a)   High pressure;
 b)   Vacuum;
 c)   Fluidization;
 d)   Any of the above.

Question 13. What is the flow rate of materials in a bucket conveyor dependent on?
 a)   Shape of the buckets;
 b)   Spacing of the buckets;
 c)   Speed of the conveyor;
 d)   All of the above.

Question 14. Which of the following is NOT an advantage of mechanical transportation?
 a)   Transportation is economical and quick;
 b)   Handling is contamination free;
 c)   No human injury;
 d)   None of the above.

Question 15. Which of the following is not a hoisting equipment with lifting gear?
 a)   Cage elevators;
 b)   Jib cranes;
 c)   Pulleys;
 d)   Troughed belts.

Question 16. What are bulk loads?
 a)   Lumps of material;
 b)   Single rigid mass;
 c)   Homogeneous particles;
 d)   Heterogeneous particles.

Question 17. Which belt conveyor prevents sliding down of material at an inclination of 55° to horizontal?
 a)   Flat belt conveyor;
 b)   Troughed belt conveyor;
 c)   Blanket belt conveyor;
 d)   Woven wire belt conveyor.

Question 18. Which of the following statements is false for troughed belt conveyors?
 a)   Troughed belt conveyors use flexible belts;
 b)   They contain five idlers;
 c)   Depth of trough decreases with increasing number of idlers;
 d)   Flexibility of belt increases as depth of trough decreases.

(a) and (b);
(b) and (c);
(c) and (d);
(d) None of the above.

Question 19. Which of the following belt conveyors has low volume carrying capacity?
a) Flat belts;
b) Troughed belts;
c) Both (a) and (b);
d) None of the above.

Question 20. Which discharge method provides only intermediate discharge for a low-speed flat belt conveyor?
a) Plow discharge;
b) Tripper discharge;
c) Both (a) and (b);
d) None of the above.

Question 21. Flight conveyors are mainly used for conveying:
a) Grains;
b) Coal;
c) Bauxite;
d) Iron ore.

Question 22. The number of cylinders in the case of a steam locomotive is:
a) One;
b) Two;
c) Four;
d) Eight.

Question 23. In the case of belt conveyors, the bearings used for return idlers are:
a) Bush bearings;
b) Split bush bearings;
c) Cast iron bearings;
d) Anti-friction bearings.

Question 24. Pneumatic conveyors are generally used for conveying:
a) Packaged goods;
b) Mineral ores;
c) Heavy goods;
d) Granular material.

Question 25. Which of the following bearings is not used in earth moving equipment?
a) Ball bearing;
b) Bush bearing;
c) Jewel bearing;
d) Needle bearing.

Question 26. Chains for material handling equipment are generally made of:
a) Cast iron;
b) Wrought iron;

c) Mild steel;
d) Carbon steel.

Question 27. Which one of the following does not fall under the category of hoisting equipment?
a) Pull lift;
b) Jack;
c) Chain hoist;
d) Dragline.

Question 28. Lift trucks are used in industries generally for the transportation of:
a) Batches of material;
b) Heavy equipment;
c) Non-ferrous materials;
d) Castings only.

ANSWERS TO MCQs

1 d; 2 d; 3 a; 4 c; 5 a; 6 c; 7 c; 8 b; 9 b; 10 d; 11 d; 12 d; 13 d; 14 d; 15 d; 16 b; 17 c; 18 c; 19 a; 20 a; 21 b; 22 b; 23 d; 24 d; 25 c; 26 d; 27 d; 28 a.

BIBLIOGRAPHY

Abdi, M.R., & Labib, A.W. (2004). Grouping and selecting products: The design key of Reconfigurable Manufacturing Systems (RMSs), *International Journal of Production Research*, 43(3), 521–546.

Battiti, R., & Tecchiolli, G. (1994). The Reactive Tabu Search, *ORSA Journal on Computing*, 6, 126–140.

Cochran, D.S., Arinez, J.F., Duda, J.W., & Linck, J. (2001). A decomposition approach for manufacturing system design, *Journal of Manufacturing Systems*, 20(6), 371–389.

ElMaraghy, H.A. (2005). Flexible and reconfigurable manufacturing systems paradigms, *International Journal of Flexible Manufacturing Systems*, 17(4), 261–276.

Herrera, F., Lozano, M., & Verdegay, J.L. (1998). Tackling real-coded Genetic Algorithms: Operators and tools for behavioral analysis, *Artificial Intelligence Review*, 12(4), 265–319.

Holland, J. (1975). *Adaptation in natural and artificial systems*. University of Michigan Press, Ann Arbor, MI.

Kimms, A. (2000). Minimal investment budgets for flow line configuration, *IIE Transactions*, 32, 287–298.

Koren, Y., Hu, S.J., & Weber, T.W. (1998). Impact of manufacturing system configuration on performance, *Annals of CIRP*, 47(1), 369–372.

Koren, Y., Heisel, U., Jovane, F., Moriwaki, T., Pritschow, G., Ulsoy, G., & Van Brussel, H. (1999). Reconfigurable manufacturing systems, *Annals of CIRP*, 48(2), 527–540.

11 Computer-Aided Process Planning (CAPP) for Automatic Manufacturing and Assembly Systems

11.1 INTRODUCTION TO PROCESS PLANNING

Process planning is a preparatory step before manufacturing which determines the sequence of operations or processes needed to produce a part or an assembly. This step is more important in job shops, where one-of-a-kind products are made or the same product is made infrequently.

11.2 OVERVIEW OF CAPP

Computer-aided process planning is the use of computer technology to aid in the process planning of a part or product in manufacturing. CAPP is the link between CAD and CAM, in that it provides for the planning of the process to be used in producing a designed part.

CAPP is a linkage between the CAD and CAM module. It provides for the planning of the process to be used in producing a designed part. Process planning is concerned with determining the sequence of individual manufacturing operations needed to produce a given part or product. The resulting operation sequence is documented on a form typically referred to as a route sheet (also called as process sheet/method sheet) containing a listing of the production operations and associated machine tools for a work part or assembly. Process planning in manufacturing also refers to the planning of use of blanks, spare parts, packaging material, user instructions (manuals), etc.

Process planning translates design information into the process steps and instructions to efficiently and effectively manufacture products. As the design process is supported by many computer-aided tools, computer-aided process planning (CAPP) has evolved to simplify and improve process planning and achieve more effective use of manufacturing resources.

DOI: 10.1201/9781003476375-11

11.3   APPROACHES TO COMPUTER-AIDED PROCESS PLANNING

There are two basic approaches to computer-aided process planning:

- Generative and automatic;
- Variant process planning.

11.3.1   GENERATIVE APPROACH

The generative approach, however, is based on generating a plan for each component without referring to existing plans. Generative-type systems are systems that perform many of the functions m a generative manner. The remaining functions are performed with the help of humans in the planning loop. Automated systems, on the other hand, completely eliminate humans from the planning process. In this approach, the computer is used in all aspects, from interpreting the design data to generating the final cutting path.

Generative type computer-aided process planning encompasses the activities and functions to prepare a detailed set of plans and instructions to produce a part. The planning begins with engineering drawings, specifications, parts or material lists and a forecast of demand. The results of the planning are:

- Routings which specify operations, operation sequences, work centers, standards, tooling and fixtures. This routing becomes a major input to the manufacturing resource planning system to define operations for production activity control purposes and defines required resources for capacity requirements planning purposes.
- Process plans typically provide more detailed, step-by-step work instructions including dimensions related to individual operations, machining parameters, set-up instructions and quality assurance checkpoints.
- Fabrication and assembly drawings to support manufacture are provided (as opposed to engineering drawings to define the part).

11.3.2   VARIANT PROCESS PLANNING

The variant approach uses computer terminology to retrieve plans for similar components using tables or look-up procedures. The process planner then edits the plan to create a "variant" to suit the specific requirements of the component being planned. Creation and modification of standard plans are the process planner's responsibility.

The variant approach to process planning was the first approach used to computerize planning techniques. Implementation of the variant approach uses group technology (GT)-based part coding and classification is used as a foundation. Individual parts are coded based upon several characteristics and attributes.

In a variant CAPP approach, a process plan for a new part is created by recalling, identifying and retrieving an existing plan for a similar part and making necessary modifications for the new part. Sometimes, the process plans are developed for parts representing a family of parts called "master parts". The similarities in design attributes and manufacturing methods are exploited for the purpose of formation of part families. A number of methods have been developed for part family formation using coding and classification schemes of group technology (GT), similarity-coefficient based algorithms and mathematical programming.

11.4 THE VARIANT PROCESS PLANNING APPROACH CAN BE REALIZED AS A FOUR-STEP PROCESS

- Definition of coding scheme;
- Grouping parts into part families;
- Development of a standard process plan;
- Retrieval and modification of standard process plan.

11.5 ADVANTAGES OF CAPP

CAPP has some important advantages over manual process planning which include:

- Reduced process planning and production lead-times;
- Faster response to engineering changes in the product;
- Greater process plan accuracy and consistency;
- Inclusion of up-to-date information in a central database;
- Improved cost estimating procedures and fewer calculation errors;
- More complete and detailed process plans;
- Improved production scheduling and capacity utilization;
- Improved ability to introduce new manufacturing technology and rapidly update process plans to utilize the improved technology.

11.6 DISADVANTAGES OF CAPP

There are a number of difficulties in achieving the goal of complete integration between various functional areas such as design, manufacturing, process planning and inspection. For example, each functional area has its own standalone relational database and associated database management system. The software and hardware capabilities among these systems pose difficulties in full integration.

11.7 FUNDAMENTAL OF AUTOMATIC PRODUCTION LINES

Automated production lines are typically used for high production of parts that require multiple processing operations. The production line itself consists of geographically dispersed workstations within the plant, which are connected by a mechanized work transport system that ferries parts from one workstation to another in a pre-defined production sequence. In cases where machining operations, such as drilling, milling and similar rotating cutter processes, are performed at particular workstations, the more accurate term to use is transfer line or transfer machine. Other potential automated production line applications include robotic spot-welding.

11.8   AUTOMATED ASSEMBLY SYSTEMS

Assembly involves the joining together of two or more separate parts to form a new entity which may be used as an assembly or sub-assembly. Automated assembly refers to the use of mechanized and automated devices to perform the various functions in an assembly line or cell. An automated assembly system performs a sequence of automated operations to combine multiple components into a single entity which can be a final product or sub-assembly. Automated assembly technology should be considered when the following conditions exist.

- High product demand;
- Stable product design;
- The assembly consists of no more than a limited number of components;
- The product is designed for automated assembly.

An automated assembly system involves less investment compared to transfer lines because:

1. Work parts produced are smaller in size compared to transfer lines.
2. Assembly operations do not have the large mechanical force and power requirement.
3. Size is smaller compared to transfer lines.

11.9   DESIGNS FOR AUTOMATED ASSEMBLY

Following are some recommendations and principles that can be applied in product design to facilitate automated assembly.

11.9.1   REDUCE THE AMOUNT OF ASSEMBLY REQUIRED

This principle can be realized during design by combining functions within the same part that were previously accomplished by separate components in the product. The use of plastic molded parts to substitute for sheet metal parts is an example of this principle. A more complex geometry molded into a plastic part might replace several metal parts. Although the plastic part may seem to be more costly, the savings in assembly time probably justify the substitution in many cases.

11.9.2   USE MODULAR DESIGN

In automated assembly, increasing the number of separate assembly steps that are done by a single automated system will result in an increase in the downtime of the system. To reduce this effect, Riley suggests that the design of the product be modular, with perhaps each module requiring a maximum of 12 or 13 parts to be assembled on a single assembly system.

11.9.3   REDUCE THE NUMBER OF FASTENERS REQUIRED

Instead of using separate screws and nuts, and similar fasteners, design the fastening mechanism into the component design using snap fits and similar features. Also, design the product modules so that several components are fastened simultaneously rather than each component being fastened separately.

11.9.4   REDUCE THE NEED FOR MULTIPLE COMPONENTS TO BE HANDLED AT ONCE

The preferred practice in automated assembly machine design is to separate the operations at different stations rather than to handle and fasten multiple components simultaneously at the same workstation. It should be noted that robotics technology is causing a rethinking of this practice, since robots can be programmed to perform more complex assembly tasks than a single station in a mechanized assembly system.

11.9.5   LIMIT THE REQUIRED DIRECTIONS OF ACCESS

This principle simply means that the number of directions in which new components are added to the existing sub-assembly should be minimized. If all of the components can be added vertically from above, this is the ideal situation. Obviously, the design of the sub-assembly module determines this.

11.9.6   REQUIRE HIGH QUALITY IN COMPONENTS

High performance of the automated assembly system requires consistently good quality of the components that are added at each workstation. Poor-quality components cause jams in the feeding and assembly mechanisms which causes downtime in the automated system.

11.9.7   IMPLEMENT HOPPER ABILITY

This is a term that is used to identify the ease with which a given component can be fed and oriented reliably for delivery from the parts hopper to the assembly work head.

11.10   TYPES OF AUTOMATED ASSEMBLY SYSTEMS

The type of work transfer system that is used in the assembly system can be:

- Continuous transfer system;
- Synchronous transfer system;
- Asynchronous transfer system;
- Stationary base part system.

THEORETICAL QUESTIONS

Question 1. Describe process planning with suitable examples.

Question 2. Describe the overview of CAPP with examples.

Question 3. Explain the types of CAPP with suitable examples.

Question 4. Describe the merits of CAPP and demerits of CAPP.

Question 5. Describe advanced manufacturing planning with examples.

Question 6. Discuss the fundamentals of automatic production lines.

Question 7. Describe the merits of automatic production lines and demerits of automatic production lines.

Question 8. Describe the applications of automatic production lines.

Question 9. Describe automated assembly systems with examples.

Question 10. Describe quantitative analysis of assembly systems with examples.

MULTIPLE CHOICE QUESTIONS

Question 1. Which system uses computers at lower-level strategies?
   a) Variant CAPP;
   b) Generative CAPP;
   c) Hybrid CAPP;
   d) All of the above.

Question 2. Which system uses computers at higher level strategies?
   a) Variant CAPP;
   b) Retrieval CAPP;
   c) Generative CAPP;
   d) All of the above.

Question 3. CAPP is called:
   a) Computer-aided product processing;
   b) Computer alternate product processing;
   c) Computer-aided processing planning;
   d) Computer alternate processing planning.

Question 4. If all the processing equipment and machines are arranged according to the sequence of operations of a product the layout is known as:
   a) Product layout;
   b) Process layout;
   c) Fixed-position layout;
   d) GT layout.

Question 5. The following type of layout is preferred to manufacture a standard product in large quantities:
   a) Product layout;
   b) Process layout;
   c) Fixed-position layout;
   d) GT layout.

Question 6. The following type of layout is preferred for low volume production of non-standard products:
    a)   Product layout;
    b)   Process layout;
    c)   Fixed-position layout;
    d)   GT layout.

Question 7. In ship manufacturing, the type of layout preferred is:
    a)   Product layout;
    b)   Process layout;
    c)   Fixed-position layout;
    d)   GT layout.

Question 8. Which of the following is not a design attribute?
    a)   Major dimensions;
    b)   Length/diameter ratio;
    c)   Tolerances;
    d)   Machine tools.

Question 9. Which of the following is not the method of part family formation?
    a)   Visual inspection method;
    b)   Automatic product sorting;
    c)   Parts classification and coding;
    d)   Production flow analysis.

Question 10. Choose the right sequence for production flow analysis:
    a)   PFA chart–data collection–sortation of process plans;
    b)   Data collection–PFA chart–sortation of process plans;
    c)   Sortation of process plans–data collection–PFA chart;
    d)   Data collection–sortation of process plans–PFA chart.

Question 11. In which type of coding structure is every digit independent?
    a)   Chain type coding structure;
    b)   Hierarchical structure;
    c)   Hybrid coding structure;
    d)   Random coding structure.

Question 12. Which of the following is not an input of process planning?
    a)   Production type data;
    b)   Raw material data;
    c)   Facilities data;
    d)   Part program data.

Question 13. Use of computers for storage and retrieval of the data for the process plans is:
    a)   Lower-level strategy;
    b)   Intermediate strategy;
    c)   Higher-level strategy;
    d)   Morden-level strategy.

Question 14. Use of computers to automatically generate process plans is:
    a)   Lower-level strategy;
    b)   Intermediate strategy;

c) Higher-level strategy;

d) Morden-level strategy.

Question 15. Which of the following produces the best quality graphics reproduction?

a) Dot metric printers;

b) Ink jet printers;

c) Laser printers;

d) Plotters.

Question 16. The user communicating directly with the computer through its peripheral devices is known as:

a) On-line processing;

b) Remote-terminal processing;

c) Batch mode processing;

d) Intelligent terminal processing.

Question 17. What key hardware item ties a CAD/ CAM system together?

a) Mouse;

b) Graphics workstation;

c) Digitizer;

d) Plotter.

Question 18. The device that assembles groups of characters into complete messages prior to their entering the CPU is called:

a) A compiler;

b) An interpreter;

c) A communications processor;

d) An editor.

Question 19. A drum timer is a good example of:

a) An output actuator;

b) An input sensor;

c) A position sensor;

d) A sequence controller.

Question 20. Flat panel displays are based on what type of technology?

a) Direct view storage tube;

b) Very large-scale integration (VLSI);

c) Solid state;

d) RGB monitor.

Question 21. Pressure sensors:

a) Use the piezoresistive effect in strain gauge sensors;

b) Use an aneroid chamber with a variable resistance output;

c) Use capacitive variations to sense pressure;

d) All of the above.

Question 22. Which of the following is a graphical input device?

a) Light pen;

b) Keyboard;

c) Mouse;

d) Track ball.

Question 23. A discrete parts process is:
    a)   The same as a continuous process except a different controller must be used;
    b)   Often encountered in manufacturing;
    c)   Normally a repetitive series of operations;
    d)   Both (b) and (c).

Question 24. Flexible manufacturing allows for:
    a)   Factory management;
    b)   Automated design;
    c)   Tool design and tool production;
    d)   Quick and inexpensive product changes.

Question 25. Productivity is defined as:
    a)   Number of items manufactured per day;
    b)   Output per man-hour of labor;
    c)   Cost per day;
    d)   Cost per unit.

Question 26. The main objective of the automation control system used in the industry is/are to:
    a)   Increase productivity;
    b)   Improve the quality of the product;
    c)   Control production costs;
    d)   All of the above.

Question 27. What is automation in the industry?
    a)   The use of control systems, such as computers or robots, and information technologies for handling different processes and machinery in an industry to replace a human being;
    b)   It is the second step beyond mechanization in the scope of industrialization;
    c)   Both of the above;
    d)   None of the above.

Question 28. Which is/are the types of automation?
    a)   Fixed automation;
    b)   Programmable automation;
    c)   Flexible automation;
    d)   All of the above.

Question 29. The machining transfer lines found in the automotive industry, automatic assembly machines and certain chemical processes are examples of:
    a)   Fixed automation;
    b)   Flexible automation;
    c)   Programmable automation;
    d)   Integrated automation.

Question 30. What is/are the basic elements of automation?
    a)   Power to accomplish the process and operate the system;
    b)   Program of instructions to direct the process;
    c)   Control system to actuate the instructions;
    d)   All of the above.

ANSWERS TO MCQ QUESTIONS

1 a; 2 c; 3 c; 4 a; 5 a; 6 b; 7 c; 8 d; 9 b; 10 d; 11 a; 12 d; 13 a; 14 c; 15 d; 16 a; 17 b; 18 c; 19 d; 20 c; 21 d; 22 a; 23 a; 24 d; 25 b; 26 d; 27 c; 28 d; 29 a; 30 d.

BIBLIOGRAPHY

Kuo, C.H. (2001). Resource allocation and performance evaluation of the Reconfigurable Manufacturing Systems, *IEEE International Conference on Systems, Man and Cybernetics*, 4, 2451–2456.

Lee, G.H. (1998). Designs of components and manufacturing systems for agile manufacturing, *International Journal of Production Research*, 36(4), 1023–1044.

Lee, G.H. (2004). Econfigurability consideration design of components and manufacturing systems, *International Journal of Advanced Manufacturing Technology*, 13(5), 376–386.

Maier-Speredelozzi, V., & Hu, S.J. (2003). Selecting manufacturing system configurations based on performance using AHP. *Technical Paper—Society of Manufacturing Engineers. MS*, no MS02179, pp 1–8.

Maier-Speredelozzi, V., Koren, Y., & Hu, S.J. (2003). Convertibility measures for manufacturing systems, *CIRP Annals*, 52(1), 367–370.

Makino, H., & Trai, T. (1994). New developments in assembly systems, *CIRP Annals,* 43(2), 501–522.

Matta, A., Tolio, T., Karaesmen, F., & Dallery, Y. (2001). An integrated approach for the configuration of automated manufacturing systems, *Robotics and Computer-Integrated Manufacturing*, 17, 19–26.

Son, S.-Y. (2000). Design principles and methodologies for reconfigurable machining systems. PhD Dissertation, University of Michigan.

Index

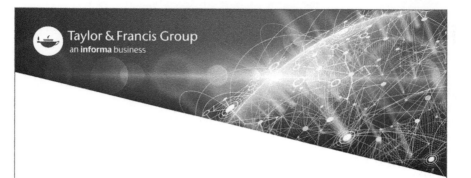